BRIDGES OVER
THE DELAWARE RIVER

Bridges over the Delaware River

A HISTORY OF CROSSINGS

~ FRANK T. DALE ~

RUTGERS UNIVERSITY PRESS
New Brunswick, New Jersey, and London

LIBRARY OF CONGRESS CATALOGING-IN-PUBLICATION DATA

Dale, Frank, 1925–
   Bridges over the Delaware River : a history of crossings / Frank T. Dale.
      p.   cm.
   Includes bibliographical references (p.   ) and index.
   ISBN 0-8135-3212-4 (cloth : alk. paper) — ISBN 0-8135-3213-2
(pbk. : alk. paper)
      1. Bridges—Delaware River (N.Y.–Del. and N.J.)—History.   I. Title.
TG23 .D35 2003
624'.2'09749—dc21
                                                                    2002011980

British Cataloging-in-Publication data for this book is available from the
British Library.

Manufactured in the United States of America

CONTENTS

PART THREE — UPRIVER

PREFACE AND
ACKNOWLEDGMENTS

I was born and raised in New Jersey, but my grandparents, both sets of them, plus aunts, uncles, and cousins, were proud but lovable Pennsylvanians. From shortly after his birth, your prospective author began crossing and re-crossing the Delaware River, usually via the old Columbia-Portland covered bridge, on visits to these loving kinfolk.

The grandparents' birth dates go back to the Civil War days, or earlier, when the river's bridges were relatively new and novel. Even in the grandparents' old age these spans, including some of the original covered structures, were part of their Delaware Valley's proud memories. I am a New Jersey Delaware Valley native son whose principal and most rewarding inheritance is this love and admiration of the Delaware bridges, something I, too, may be able to pass on. I hope so.

Other sources of historical bridge information are listed in this book's bibliography, but there were more sources than that. Old newspaper clippings and family letters in attics and basements, oral tales passed down from earlier and recent generations, several hard-working and, for the most part, generous historical societies, and finally, a large, but at first hard to find, group of river valley photo collectors. Their photos, copies of them, were slowly but gratefully received and put to use. At the risk of omitting the name of one or more of these many and varied helpers, I would like to say, "Thank you," once again, to: Mary Curtis; Frank Schwarz; Lester Wellington; the Wayne County Historical Society and especially Anne; the Equinunk Historical Society; Paula Valentine, Dot Moon, and anonymous helpers, all from the National Park Service; the staff at the Honesdale Library; the staff at the Narrowsburg Public Library; Mr. and Mrs. Bob White; the Pike County Historical Society and Laurie; the Delaware County Historical Society; the Sullivan County Historical Society; George Fluhr; the Delaware River Joint Toll Bridge Commission, with special thanks to Albert Picco, George Alexandridis, and Jim Gambino; Robert Longcore, Nan Horsfield, and Nellie Stires Howell, all of the Sussex County Historical Society; Bill Lifer; Alicia Batko; Peter Osbourne and the Minisink Valley Historical Society; Roxanne of the Hunterdon County Historical Society; Emily Hallock of the Tuston-Cochecton Historical Society; the Bucks County Historical Society, with Cynthia and Donna; the Mercer Museum; John Bade; Tom Drake; the Doylestown Public Library; Carol Phillips and John Perkins of Dingman's Ferry; the Horseheads Historical Society; Sharon and Barbara of the Marx Room of the Easton Public Library; the Hackettstown Historical Society; the Port Jervis Library; Len Peck; Janet at the Warren County Library; Morris Scott; Ruth Hutchinson Gommoll; and, last but not least, Edwin Harrington. Thanks again to all of you.

BRIDGES OVER
THE DELAWARE RIVER

Introduction

Ferries on the Delaware River were there first and, at their beginning, were adequate. But as traffic between the states increased, problems arose. Often the ferries were not large enough to carry a traveler and his coach and horse and certainly not large enough to handle two or three at a time. The same was true of wagons carrying goods.

One of the early travelers was Robert Hunter, an Englishman, who wrote in his travel diary about crossing the river on a small ferry called a wherry, at Trenton; "We crossed the Delaware on a wherry and remained in our coach the whole time. The scow was very small and it was really dangerous. There was only an inch between us and eternity."

By the end of the eighteenth century, the busier crossings—at Easton, Trenton, or Stockton, for example—had frequent lines of travelers and wagons waiting their turn. Another problem was that ferries were slowed down or stopped entirely at times of high water or when the river froze over, or at spring thaw when the river was filled with large chunks of ice. And as the other traffic on the river increased, especially the huge, unstoppable timber rafts, crossing the river by ferry became extremely hazardous. Horses seemed to be especially skittish on the water and, in a crisis, would sometimes spook and jump overboard, often with the carriage attached. Frequently, the poor beast would drown before it could be cut loose.

After the Revolutionary War, when our nation's capital was located in Philadelphia, our roads and ferries were busy with New England and Middle Atlantic states' congressmen and others on their way to the city. This was still true when the capital was moved to Washington, D.C.

Ferries on the Delaware also carried a great deal of commercial traffic going southward or westward. The crossing at Easton was referred to as the New York–Harrisburg Crossing, indicating that the ferry here served travelers going to western Pennsylvania, and beyond. The two most frequently used river crossings for this traffic were at Trenton and Easton, and these crossings were where the first bridges over the Delaware were built. By the mid-nineteenth century many of the river's ferries were replaced by bridges. Only a few ferries survived into the twentieth century. This story is mostly about those early bridges over the Delaware, some of which still stand and serve.

The franchise to operate a ferry in colonial days was a royal grant, awarded to private individuals; a ferry could not be operated without such a grant. This protected the ferry from competition. The reason for giving such a grant was often to reward loyalty or service to the crown, and the idea was that the ferry owner would charge a toll and make money. So the precedent was established, early, that there would be a toll for crossing the river. This carried over; the bridges, like the ferries, could be built only with the approval of the legislatures of the states involved, but would be privately owned, usually by a corporation of local stockholders. This organization was expected to charge a toll and to make a profit, but also to maintain and repair the bridge. If the bridge was damaged or destroyed in a flood, the corporation was expected to repair it or build a new one, and it usually did one or the other—these bridges were money-makers.

Although local toll collectors usually allowed churchgoers and schoolchildren free passage over the bridge, it wasn't until the twentieth century that people and politicians began to talk about totally toll-free passage over a river. Finally, in the 1920s, the states bordering the Delaware River established a Joint Commission for the Elimination of Toll Bridges. Its purpose was to purchase bridges from the private owners, at a price agreeable to the stockholders, and then make each bridge a "free bridge," one that did not charge a toll. A few bridges would be built on major highways and would continue to charge tolls, but for many shoppers and workers who lived along the river and crossed it often, tolls were eliminated. Today, only one privately owned bridge still crosses the river, and it still charges a toll. This bridge, between Dingman's Ferry, Pennsylvania, and Layton, New Jersey, is called, unofficially, the Dingman's Ferry Bridge.

Now, for more about the history of the bridges on our Delaware, including that unique Dingman's span, read on.

PART ONE

Downriver

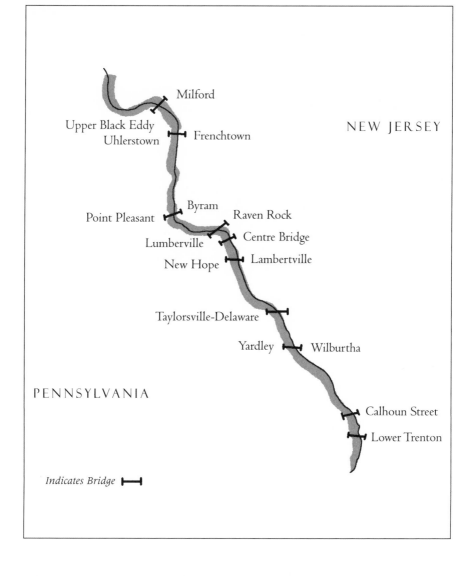

Milford

Upper Black Eddy

Uhlerstown   Frenchtown

NEW JERSEY

Byram

Point Pleasant   Raven Rock

Lumberville   Centre Bridge

New Hope   Lambertville

Taylorsville-Delaware

Yardley   Wilburtha

PENNSYLVANIA

Calhoun Street

Lower Trenton

Indicates Bridge

ONE

The Lower Trenton Bridge, 1806

The Delaware River at Trenton enters into a long series of sharp, downriver rapids called Trenton Falls; and below these rapids the river enters into the deeper and wider tidewater section, on its way to Philadelphia and then the ocean. Trenton was the last stop before the river entered this ocean-going section and as such was, early on, a busy area.

By the mid-1600s there seems to have been a ferry crossing the river at Trenton, one at the base of the Falls, called Lower Ferry. And soon there was a second ferry, located about a mile upriver at the beginning of the downriver Falls and called Upper Ferry. Pennsylvania historian B. F. Fackenthal indicates that the lower ferry was in business first, in the mid-1600s, and that both ferries traveled between what is today called Morrisville, Pennsylvania, and Trenton, New Jersey.

The first ferry owner whose name we know was James Trent, and he was granted, in 1726, both of the New Jersey ferry locations, Lower and Upper. This soon changed, with separate ownership of the Lower ferry. In 1745, Thomas Hooten was operating this ferry and, in 1753, Robert Hooper had taken his place; by 1770, William Richards was operating it.

1 — The first Lower Trenton Bridge and the first Delaware River bridge. Photo courtesy of the author.

The very first bridge built across the Delaware River spanned the Lower Level area at Trenton, which had become the busiest ferry crossing on the river. This bridge connected Trenton with the Borough of Morrisville, in Pennsylvania. The large, wooden span was 1,008 feet long, and, although its construction began after the upriver Easton-Phillipsburg Bridge was underway, the Trenton Bridge opened for business first, on January 30, 1806. The Easton span didn't open until November of 1806, but it lived a much longer life than the Lower Trenton structure. And the large Trenton span cost $180,000; by comparison, the Easton Bridge, spanning the narrower river, well upstream, cost only $65,000.

The designer and builder of the Trenton Bridge was Theodore Burr, the foremost American bridge builder of his time. In 1804, he completed a bridge in Waterford, New York, which was the first to span the Hudson River. His skill and experience explain the relatively long and trouble-free life of most of his bridges. The first president of the privately owned bridge company at Trenton was John Beatty, bearer of a name that would make a mark in New Jersey history.

The bridge deck was 31 feet wide, which provided two 11-foot-wide roadways for horse and wagon, as well as two 4-foot walkways, one on each side of the deck, for pedestrians. It was a covered bridge, but as it turned out, a unique one; it had a cedar-shingled roof and partially enclosed sides consisting of a wall, 4 feet in height, all along the outside of the bridge to protect wagons and pedestrians. The piers and abutments that supported the bridge were designed to be high enough to avoid all known flood levels, but during construction, the river flooded and threatened the incomplete bridge. Burr promptly decided to raise the piers and abutments a few more feet. This bridge would thus survive all future floods.

During the river's first big flood after the appearance of bridges, in January of 1841, six other, newer bridges over the Delaware were swept away, but the Trenton span stayed high and dry, and undisturbed. The only other bridge to survive this flood was the Palmer Bridge at Easton-Phillipsburg, the second river span to be built. These two bridges that survived this calamity were the river's oldest.

Theodore Burr became famous for his bridge design, but he fell into debt, nevertheless, and when he died in 1822, suddenly and for unknown reasons, his family could not afford funeral expenses. He was buried in an unmarked and now unknown grave.

In 1842, Burr's Trenton structure became the first bridge in the United States to be used for interstate railroad traffic. One of the two horse-and-wagon lanes was laid with track and converted to use only for railroad traffic; the remaining horse-and-wagon lane was subjected to something now called traffic jams. Finally, in 1848, after six years of confusion, the bridge was widened enough to give a separate lane to the trains and return the original train lane to the horse-and-wagon combination. The first impression made by rail traffic on the bridge was good; historians Barber and Howe, in *Historical Collections of New Jersey,* referred to the bridge as "one of the finest specimens of bridge architecture in the world. It withstood the great flood of 1841 while most other bridges were swept away. It is crossed by the Philadelphia and Trenton Railroad." This last sentence points to the cause of the short life of this fine structure.

Even the early trains were heavy and a strain on the wooden bridge. The first engine in use in the United States, which arrived from England in 1829, weighed 3.5 tons without any freight cars attached and traveled at only about 20 miles an hour; by 1880 a passenger locomotive weighed about 70 tons and traveled at 60 miles an hour. The relatively early demise of the Trenton Bridge was undoubtedly caused by the heavy railroad trains and traffic, for which this wooden bridge was not designed.

The bridge was busy, even overworked, during the next fifteen years, by heavy railroad traffic. Then a fire broke out on the Jersey end of the span, caused by sparks from a locomotive's engine, and the entire bridge was threatened but survived. The bridge company continued the profitable rail traffic, however, but removed the wooden-shingled roof and the other fire-prone wooden enclosures on the bridge deck.

So the bridge worked hard and paid for itself many times over but, by the middle of the century, it was weakened by old age and the hard labor demanded by its unique customer, the railroad. Then, in 1874 the railroad decided to build a set of two tracks separate from this original bridge. They extended the old bridge's piers and abutments on the downriver side and laid new railroad trackage on them. The original bridge went back exclusively to horse-

and-wagon customers. But the damage had been done; just a year later, in 1875, the old and obsolete vehicular bridge, worn out by train traffic, was torn down and a new iron bridge was erected nearby in its place. The original Delaware River bridge, built in 1806, was no more. The nonrailroad Easton-Phillipsburg Bridge, built the same year as the Trenton Bridge, would last another quarter of a century.

This iron bridge built in Trenton and opened for business in 1876 was for vehicular traffic only. The new structure was designed by Joseph Wilson and built by the Keystone Bridge Company of Pittsburgh. Bridge owners had learned a lesson; this bridge would remain a vehicular span.

The railroad, in 1892, constructed a new steel bridge for itself crossing the Delaware in this Lower Trenton area. It contained four sets of tracks. And in 1898 the railroad span built in 1874 on the old bridge was torn down and replaced by another new steel four-track railroad bridge at the same location. Both of these new railroad bridges were constructed by the American Bridge Company.

Finally, in 1908, the Pennsylvania Railroad built a new and beautiful stone arch railroad bridge just downriver of its older bridges. The company then took down the bridges built in 1892 and 1898 and shipped these nearly new structures to Washington, D.C., where they would carry traffic over the Potomac River to the nation's capital.

The iron bridge built in 1876 was used for foot and carriage traffic and then auto traffic, as well. And it remained a profitable and relatively trouble-free toll bridge. Then, on July 12, 1918— World War I was underway and bridges were militarily vital—this bridge was taken over by the Joint Commission for Eliminating Toll Bridges at a rather attractive price of $240,000. The bridge's tolls were eliminated at once. The bridge was older and overworked, with the coming of motor vehicles and heavily loaded trucks; it became increasingly unsuitable for modern vehicular

2 — Ice gorge at the new Lower Trenton Bridge, looking downriver, March 1934. Photo courtesy of the Delaware River Joint Toll Bridge Commission.

traffic. Finally, in 1928 this toll-free bridge was torn down and a new, more sturdy and up-to-date bridge was constructed. Only the original piers and abutments could be reused. The bridge's approaches were reconstructed at the same time. The total cost for all this work was $650,000.

The Lower Trenton Bridge has led a productive and peaceful life since that time, devoid of flood damage and owner shenanigans. Its ancestor bridge was the first one over the Delaware and from it, future bridge owners and governmental agencies learned many lessons. And so did the railroads. This provided the blueprint and rulebook for future river spans.

TWO

The Centre Bridge at Stockton, 1814

Prior to the construction of a bridge over the Delaware River in what is now Stockton, in Hunterdon County, a ferry had been in business since about 1700. This ferry was owned and operated by John Reading and was called Reading's Ferry. The little settlements on both the river's banks served by his ferry were also known as Reading's Ferry. John Reading was a prominent citizen in the area until his death in 1717.

For several years the ferry ceased operation, but in 1731, John Reading's son-in-law, Captain Daniel Howell, took over the ferry and operated it until 1770, when ownership passed to another family member, Joseph Howell. It was known during this period as Howell's Ferry. The last owner of the New Jersey ferry was, briefly, a Mr. Robinson; it was then called Robinson's Ferry.

The ferry's Pennsylvania landing was in other hands by this time. It was owned and operated by a prominent businessman, William Mitchell, and was, of course, called Mitchell's Ferry. It operated profitably until the bridge company bought out both ferry owners and opened its bridge in 1813.

When the Centre Bridge Company bought the ferries' properties on both sides of the river, it paid for them by giving to each of the two ferry owners forty shares of common stock in the new bridge corporation. Previously depressed, William Mitchell now became an active bridge owner.

The bridge at Stockton was the first Hunterdon County structure to be erected across our Delaware River. It was built and operated by a privately owned firm, the Centre Bridge Company, and was named, officially, "Centre Bridge." Although this bridge, and all the other early ones across the Delaware, were privately owned, money-making operations, approval first had to be given by both the Pennsylvania and New Jersey state governments. The structure was paid for with funds generated by the sale of stock in this privately owned corporation.

The stock certificates were sold locally, primarily, and were often purchased by older, retired citizens. Shares in this bridge corporation were acquired by that Pennsylvania ferryman, now retired, William Mitchell, who hoped this would compensate him for his financial loss caused by this bridge. The first meeting of the stockholders of the bridge corporation, before construction began, had been held on October 21, 1812, and a board of directors was chosen. This group, at this early meeting, acquired the land owned by the two ferrymen for the bridge site. At the next meeting, held on Christmas Eve, 1812, the board of directors hired as the construction supervisor Captain Peleg Kingsley, a Northampton, Massachusetts, bridge engineer, and Benjamin Lord as his assistant. They also hired the bridge's first toll collector, John Abell, whose employment would begin on January 1, 1814.

The new span opened for business in the spring of 1814. It was called Centre Bridge because of its location halfway between the first Delaware bridge, which opened downriver at Trenton in January of 1806, and the second, upriver at Phillipsburg, which also opened in 1806, but in November. The ferry port that existed on the Pennsylvania riverbank opposite Stockton adopted the name "Centre Bridge" when the bridge replaced the ferry. The ferry at

this bridge site, although out of business, would reappear occasionally when the bridge was temporarily incapacitated because of flood damage or defective construction.

The Centre Bridge, like the two earlier bridges upriver and downriver, was a wooden covered bridge. Steel would come later. During the winter of 1816–17, a fine stone tollhouse was built as home and office for toll collector John Abell and family at the New Jersey end of the bridge in Stockton. This structure was built by stockholder and former ferryman William Mitchell. Tolls were collected prior to the erection of this building, but now life would be more comfortable for the toll collector.

It soon became the practice of toll collectors to permit free crossings to churchgoers on Sunday, or morticians on funeral day, and others. This was a custom first instituted by ferrymen. It was appreciated by all parties concerned until it was discovered that bootleggers dressed as morticians sometimes crossed the river driving a hearse . . . filled with bottles instead of bodies.

This new covered bridge underwent repairs and redesign as early as 1829 for problems caused by faults in the initial design, and it was exposed to flood and fire damage as well. The great flood of January 8, 1841, was an ice-filled destroyer that carried away three of the bridge's six spans, two of the bridge's piers, and damaged the bridge's fine tollhouse. All of this destruction occurred on the Jersey side.

A local historian, Hubert Schmidt, in *Rural Hunterdon*, tells us of George B. Fell, who was working as a substitute toll collector on the bridge the day of the flood. Ignoring the danger of the rising water, George Fell walked out on the bridge to observe the conditions, and the wooden structure collapsed beneath him. He clung to a floating plank and then managed to board the bridge roof as it passed him. He was carried downriver in the turbulent and frigid winter water. He lay flat to get under the New Hope span, and just as he passed under the Yardleyville Bridge it collapsed, barely missing him. He continued his trip, unable to be rescued until he got within three miles of Trenton, where he managed to float into a backwater area and was finally saved. He survived, barely, and immediately returned upriver, where he was received as a hero.

This flood, named the "Bridges Freshet," destroyed, totally, six bridges on the river, and seriously damaged all of the river's other spans that existed at this time. The Centre Bridge was promptly rebuilt by contractor Cortland Yardley and raised six feet higher than it had been in the 1841 flood. The new structure was stronger but had few piers, to facilitate the passage of timber rafts. The repaired bridge was back in business before the year was over, literally better than ever. As business increased, thanks to the fine bridge, and the population in this settlement grew, it was decided, in 1851, that it had achieved the status of a village and needed a post office. The postal department gave this village the official name of Stockton.

In 1852, the state of Pennsylvania, and then the state of New Jersey, approved the addition to this bridge of a rail line to carry mined material from a quarry in nearby Solebury, Pennsylvania, to New Jersey. This business venture never materialized and the work was never undertaken. This was probably a blessing in disguise, judging from the early experiences of the Lower Trenton Bridge.

The bridge received minor damage in a major flood that occurred during the Civil War, on June 6, 1862, just a year before the great battle at nearby Gettysburg. The lightly damaged Centre Bridge remained in service, however, and was quickly repaired. It was the only river span between Easton and Trenton to remain in operation through this flood.

At about this same time a private bridge, that is, one built only for private, not public, use, was constructed nearby to span part of the Delaware from Solebury, Pennsylvania, to the 121-acre Hendricks Island, just a quarter of a mile upriver from the big Centre Bridge at Stockton. Local folks called it the "Roebling Bridge." It was a medium-sized suspension span, Roebling's specialty, and Roebling's factory, by this time, was just a few miles downriver at Trenton, but whether or not Roebling did the erection or whether it was simply a copy, is unknown. This span survived the major floods of 1862 and 1903, but needed major repairs after the latter. It was finally destroyed by the flood of 1955 and not rebuilt.

The first river catastrophe of the twentieth century was the flood of 1903, flowing from an ice-filled river in October. It was, at that time, the worst Delaware flood on record, but, miraculously, the rebuilt Centre Bridge survived, virtually undamaged. Since the Centre Bridge's deck level had been raised six feet after the 1841 calamity, it had survived, nicely, all the floods to which it had been exposed. Every other Delaware bridge between Phillipsburg and Trenton was hard-hit in the flood of 1903; many were destroyed.

The Centre Bridge was threatened again just two years later. At midnight, January 26, 1905, a fire broke out at the Davison

4 — Centre Bridge after the 1923 fire. Photo courtesy of the Delaware River Joint Toll Bridge Commission.

General Store on Bridge Street, adjacent to the Centre Bridge. The store was quickly consumed, as well as nine nearby homes. The outside temperature at the time was zero, and over a foot of snow was on the village streets. Stockton didn't have a fire company but Lambertville citizens did, and they sent it by train to burning Stockton. The fierce fire was finally extinguished at the very entrance to the old Centre Bridge.

The "Roaring Twenties" may have been boom times for some businesses in this area, but it was a hard decade for our river bridges. The now old and wooden, covered Centre Bridge was damaged, but not destroyed, by fire on May 26, 1923. Again, two months later, on July 19, it was badly burned. Finally, just three days after that, on July 22, the bridge was struck by lightning in the early evening. The wooden structure went up in flames and was quickly and totally destroyed. The only remnants were the bridge's charred piers, standing desolately in the river . . . and the un-damaged tollbooth. No attempt was made to rebuild the bridge, and ferries were back in business in Stockton.

This ended the investment value of this bridge, which had been in existence since 1814, and had led, at best, a precarious life. It had paid only small dividends from its birth until 1829, and after that, for the next forty-four years, paid nothing at all. Shareholders were undoubtedly anxious to sell their stock, but who would buy it? Then, in 1873 dividends began again to reach the patient stock-holders, and even rose after 1885. Then in 1923 the bridge was totally destroyed by fire. Dividends were never paid again, and when the states bought the bridge remnants for a meager $10,000, there wasn't enough money left to do much more than finance a farewell party for these unfortunate stockholders.

Two years after the fire, in November of 1925, the New Jersey–Pennsylvania Joint Commission, an agency set up to take over ownership of Delaware bridges, purchased this bridge rem-nant from the discouraged private corporation, Centre Bridge Company. The Joint Commission offered, and the Centre Bridge owners accepted, the small sum of $10,000. The Commission promptly rebuilt the ruined piers and abutments, encasing them in concrete, and then erected a bridge of fire-proof steel girders with an additional section to carry the bridge over the adjacent Delaware and Raritan Canal bed. A total of 976 tons of steel went into the construction. And it was no longer a covered bridge. The level of the bridge's deck was raised to 33 feet and 8 inches above normal water levels to avoid future floodwaters, and as of this date, 2003, it has succeeded. This beautiful, steel structure was opened for business by the Joint Commission on July 16, 1927.

The life of the new structure has been a pleasant one for the most part. The old stone tollhouse built in 1816 that had survived the damage that destroyed the bridge, including the fire of 1923, was torn down, on general principles, in 1952. At about the same

5 — Centre Bridge at Stockton, N.J., taken from Pennsylvania in 2002. Photo courtesy of the author.

time, officials built a modern officers' shelter elsewhere on bridge property, but this new building's foundation soon collapsed and the structure was dismantled. Its replacement was then erected on the tollhouse's almost two-hundred-year-old foundation and stands there nicely to this day.

The bridge is maintained regularly and has survived all the floodwaters, including the worst yet to come to the Delaware valley, the flood of 1955. During this flood, the Delaware and Raritan Canal running at river's edge in Stockton was torn up, and homes along the river's edge were flooded to the second floor. But the new steel structure with its increased elevation survived and lost no time. Mother Nature has not yet been able to raise water to the higher bridge deck.

This Gibraltar-like structure stands today, the pride and joy of valley residents and river crossers alike. And now it's toll-free! So try it out; it's well worth a visit.

THREE

The New Hope–Lambertville Bridge, 1814

The second Hunterdon County bridge to span the Delaware was the New Hope–Lambertville Bridge. This structure was completed in the same year as the Centre Bridge at Stockton but a few months later, in September of 1814. It was built about three miles down-river from Stockton and located, like the Centre Bridge, in an area already busy as a ferry crossing.

A Pennsylvanian, John Wells, was, in 1722, the first licensed ferry owner operating between New Hope and what became Lambertville. And for many years before that date, he operated his ferry here without a license. The dangerous rapids just below Lambertville got the name "Wells Falls" from him. Wells also owned a fine inn at New Hope. The first ferry to operate from the New Jersey side in this area was established by Emmanuel Coryell in 1726 and then operated by his son, John. The Wells Pennsylvania-based ferry was bought out by Benjamin Canby in 1745. Canby died and left the ferry to his wife. She soon also died, in 1760. Her family then sold the business to John Coryell, who already owned the ferry across the river. Coryell operated his enlarged ferry during the Revolutionary War, and the small but growing New Jersey village became known as "Coryell's Ferry." During a portion of the war,

General George Washington and his troops occupied the Pennsylvania side of the Delaware, including the Coryell's Ferry landing, and briefly took over all the river's boats, including the ferries, so the British army could not cross over and attack. Shortly afterward, on December 26, 1776, Washington's army attacked the British a few miles downriver, at Trenton, with great success. The Continental Army crossed the Delaware at Coryell's on four occasions during the war.

After this wartime experience, John Coryell got into another business, a public house of entertainment in his hometown of Solebury, on the Pennsylvania shore of the Delaware. His son, Abraham, who moved to Hunterdon County, took his father's place with the ferry and operated it from the Jersey side for a time. The last owner of this ferry, before the bridge replaced it, was Joseph Lambert, a member of another prominent family in the area. It was, of course, renamed Lambert's Ferry. The village on the Jersey shore was named Lambertville in honor of another family member, Senator John Lambert. When the post office opened here in 1814, the first postmaster was the Senator's nephew, namesake, and a war veteran, Captain John Lambert. By this time, the number of river crossers had outgrown what the ferryboats could handle with promptness. The bridge was welcomed.

The bridge-building organization, the New Hope–Delaware Bridge Company, was another privately owned corporation that would build the bridge and operate it as a business afterward, with some government regulation. The bridge company hired for its chief engineer Lewis Wernwag, a man with an excellent reputation in this field.

The bridge was a wooden, covered structure, which had a length of 1,050 feet and 6 inches, making it a hundred feet longer, and more expensive, than the Centre Bridge. Chief engineer Wernwag used iron bracings extensively in the structure, a practice that was seventy-five years ahead of time but also ran up the bridge's cost. The final cost for this construction was $68,000. The bridge opened for business September 13, 1814, just six months after the Centre Bridge.

The New Hope–Delaware Bridge Company was also given banking privileges by the state governments. This was supposed to be a solution to the lack of banks in the area but instead proved to be harmful to the bridge corporation's primary function, efficient river crossing. The bridge company was soon getting involved in shaky business loans and sometimes losing the money collected as bridge tolls. So, although business in both Lambertville and New Hope was thriving and both villages were growing, the bridge itself was mismanaged and soon in bad financial shape. This situation, combined with the structural shortcomings of the bridge that called for immediate and expensive repairs, caused serious problems. And they were made worse by a depression, the Panic of 1837, and finally, by the flood of 1841. An eyewitness in Elmer Roberson's *Historical Collections* tells of the last calamity:

> At about half-past ten o'clock fears began to be entertained for the safety of our bridge. The river was then nearly up to the bridge. The ice and drift stuff increased and struck the piers and timbers of the bridge with tremendous force. Large coal boats, heavy saw logs, and cakes of ice had forced apart one or two of the piers on the Jersey side. Then, about 11 o'clock we heard the astounding cry, from many voices, that Centre Bridge was coming. All eyes were fixed upon two massive pieces of the bridge which were seen floating down. One of the pieces struck about midway with an awful crash and took with it one of the arches of our bridge. The other soon followed and took another arch on the Jersey side. The Jersey pier soon gave way and the third bridge arch followed. Thus, one-half of this noble structure, which had withstood freshets for nearly thirty years, has been suddenly carried away.

6 — New Hope–Lambertville covered bridge, built in 1842 after the original bridge was destroyed. This bridge lasted until the flood of 1903. Photo courtesy of Robert Longcore.

Unknown to this onlooker was the fact that the Centre Bridge's toll-taker, George B. Fell, was riding downriver on a piece of his bridge's wreckage and just barely avoided certain death from the collapse of the New Hope–Lambertville Bridge after it was struck by a large section of the Centre Bridge.

A local ferry took over while an almost totally new bridge, still a wooden covered bridge, was erected at a cost of $40,000. This unexpected financial loss destroyed the bridge corporation. First, it lost its banking privileges, and then, in 1853, it also forfeited its ownership of this new bridge. Possession of this financially distressed bridge was then taken over by two men, John Michener and James Gordon, but they soon sold their bridge to Samuel Grant of Philadelphia as the sole owner. In 1855 the first bank in the county opened up in Flemington and took future bridge companies permanently out of the banking business. Private individual ownership of this bridge lasted over thirty years. Then, on May 7,

1887, a new bridge company came into existence, eager to own the bridge and with available capital of $30,000. Under the competent new ownership, this capital increased to $70,000 by 1898. At last the bridge was in good and prosperous hands . . . and then Mother Nature took over.

In December of 1901, raging floodwater came within 9 inches of the bridge's deck. And just a year later the "Ice Freshet" raised the river 20 feet at Lambertville. The bridge survived both events.

Then, just ten months later, on October 10, 1903, this bridge was swept away in what at the time was the worst Delaware flood in recorded history; even the bridge's masonry substructure was carried away. The new bridge company immediately hired a steam launch with a capacity of twenty-five passengers to operate between Lambertville and New Hope. This got Pennsylvania commuters to the Jersey railroad station at Lambertville and home again.

The new replacement bridge, the present structure, was constructed in 1904. By this time steel was being used

7 — New Hope–Lambertville wagon and trolley bridge, opened in 1904. Photo from the Collections of the Hunterdon County Historical Society.

in major construction and this bridge was such a structure. It consisted of six all-steel spans and cost $63,818. This was less than the cost of the original crude wooden structure built back in 1814. And this new bridge became even more important with the development, in New Hope in 1906, of the first artificial ice plant in the Delaware Valley. Vast quantities of this new product were bridged across the river to Lambertville and then shipped by train to the big cities.

This New Hope–Lambertville Bridge was the first bridge purchased in Hunterdon County by the New Jersey–Pennsylvania Joint Commission for the Elimination of Toll Bridges; the sale date was December 31, 1919, and the price was $225,000. A handsome profit was made by the previous owners, over the cost of this bridge just fifteen years earlier. And on this same day the bridge was freed of all toll charges. Now this bridge was a

8 — Lambertville Bridge and former tollbooth from the New Jersey side, 1924. Photo courtesy of the Delaware River Joint Toll Bridge Commission.

New Jersey–Pennsylvania government-owned agency.

9 — New Hope–Lambertville Bridge in 2002. Built in 1904 and rebuilt in 1955. Photo courtesy of the author.

Repairs and updating, under this new ownership, took place through the years. The wooden deck on the bridge was replaced with steel grating in 1947, and at the same time the trusses were strengthened to increase the bridge's capacity to handle modern and heavier cars and trucks. Later, the bridge's motor vehicle approaches were all repaved and so was the still-used pedestrian walkway on the bridge.

Then, in the record-breaking flood of August 19, 1955, the secure life of the new bridge was threatened. Span number 2 was completely shattered and then swept away by the high water that was filled with floating and bobbing houses, trees, barns, and wagons. About 220 families living along the river were evacuated and 133 homes and five businesses had water on the first floor. And just as serious was the destruction of the town's sewage disposal plant, just built and scheduled to open for the first time in about two weeks. Local citizens would have to live a little longer with their more primitive cesspools. The bridge was closed the next day and stayed closed for a month and a half, by which time emergency repairs, with day-and-night labor, got the bridge back in operation. This work was done under the supervision of the Army Corps of Engineers.

Since the flood of 1955, the life of the bridge has been relatively uneventful. The bridge is regularly given maintenance updates and the minor problems that occur are promptly taken care of without any inconvenience to users. The bridge is also regularly cleaned and painted. And every few years, the structure is subjected to an in-depth inspection and structural analysis. So far everything looks excellent.

These fine old structures have become integral parts of Delaware River life and have an exciting past of their own. Let's hope they stay with us.

FOUR

The Taylorsville-Delaware Bridge, Washington's Crossing, 1834

The river crossing at Washington's Crossing not only served Washington and his army, but many others as well, mostly as a ferry in the early, pre-bridge days. The village on the river on the Pennsylvania side was called Taylorsville and the ferry, Taylor's Ferry, after the owner at the time. The first ferryman, however, was Henry Baker, who operated Baker's Ferry here in the late 1600s. It was in his family about one hundred years; the last family member who owned the business was Samuel, Henry's grandson.

In December 1774, the ferry had a new owner, Samuel McKonkey. He was the man who made his mark in history as the ferry owner when General Washington and his bedraggled army recrossed the river from Pennsylvania to attack the British at nearby Trenton. They scored a major, and much needed victory here on Christmas night of 1776. The army had crossed the river on large Durham boats, not McKonkey's ferryboat, but had utilized the ferry's docks on both shores. This was McKonkey's claim to fame as a ferryman. He didn't remain long as the owner, however. On May 21, 1777, Samuel McKonkey sold his ferry to a Pennsylvanian, Benjamin Taylor, and it became known, officially, in

Pennsylvania, as Taylor's Ferry. A ferry that had its home base on the Jersey side had a different owner and a different name; it was referred to, in Jersey, as Tomlinson's Ferry. When a post office was established on the Pennsylvania shore here in 1829, the village was christened Taylorsville, home of Taylor's Ferry. However, Samuel McKonkey, who owned the ferry for only three years, became the local hero, ranking right up there beside George Washington.

As early as 1831, the legislators of New Jersey and Pennsylvania authorized the creation of the Taylorsville-Delaware Bridge Company to construct a bridge at this busy Taylor's Ferry crossing. Certain standards were set by both states to avoid future problems. The states authorized that the new company could get money from selling stock, $20,000 worth, but only when $15,000 worth of stock was sold could the company start building its bridge. The two state governments insisted that the bridge company set aside a percentage of its income to pay for maintenance and repair of old age deterioration. Finally, the company was prohibited from engaging in any banking operations. Under these regulations the Taylorsville-Delaware Bridge Company went into action and by 1834, the bridge was opened for business and the ferries went into retirement.

The bridge was typical of the time, a wooden covered bridge with six spans that stretched across the 875 feet of the river's width here. Its elevation was 23 feet above normal water level. The bridge seemed to please local folk with the exception of the retired ferry operators. But not for long. After just seven years of satisfactory operation, the disastrous flood of January 8, 1841, struck the area and carried away the whole bridge. This was the worst river flood of the century. The damage was increased by the pieces of destroyed upriver bridges that served as battering rams; every Hunterdon County bridge was badly damaged or destroyed. The overwhelmed bridge company had not yet set aside enough toll money

to replace the whole structure. However, the firm's ambitious executives borrowed the full amount needed with promissory notes. The bridge was totally rebuilt, and this indebtedness was eventually paid off.

The new structure, still a wooden covered bridge, seemed superior to the original. It withstood several floods and lasted many more years than the previous bridge. It survived the nineteenth century nicely, but not by many years. In the flood of October 10, 1903, the entire superstructure was, again, carried away; only the bridge's piers and abutments remained standing. Because of its earlier indebtedness, the bridge company now had a major financial problem and no solution to it. The "Taylorsville-Delaware Company" went out of business and was replaced by another corporation, the "Taylorsville Delaware Bridge and Washington's

11 — Taylorsville, or Washington's Crossing, Bridge in the flood of 1903. Photo courtesy of William Lifer.

Crossing Delaware Bridge Company." The new corporation was able to sell $27,750 worth of stock, which was enough, just barely, to replace the previous wooden covered bridge with a modern steel structure; that replacement cost $26,775.25. This six-span steel bridge started a new life under a new owner and functioned well. It seemed capable now of surviving any flood that might occur. Nevertheless, when this bridge's stockholders got an opportunity to sell their bridge, they didn't refuse.

The bridge was purchased by the Joint Commission for the Elimination of Toll Bridges on April 25, 1922, for an irresistible sum of $40,000, and all tolls were, of course, eliminated. Considerable updating of the bridge took place immediately in 1922–23, with additional work done in 1926, 1947, and 1951. This new Joint Commission structure did an exemplary job until the arrival of the terrible flood of 1955. This deadly catastrophe of August 19 inflicted considerable damage to the new steel structure. Floating debris filled the river and moved rapidly downstream. Whole trees, complete houses, steel barrels, and even railroad freight cars smashed against this bridge and badly damaged all six spans. Little of the bridge, except the underwater piers and abutments, survived unscathed. When the floodwaters subsided, the major repair work got underway. This was performed by the Army Corps of Engineers, and their Bailey trusses supported much of the bridge temporarily while new permanent trusses were installed. And to make this calamity somewhat more bearable for the Joint Commission, the complete cost of repairs for this Washington's Crossing landmark was paid for by the Army Corps of Engineers. The almost-new bridge was open for business again on November 17, 1955; only the old piers and abutments could trace their beginnings back to the original structure of 1834.

The most recent repairs were reported in 1994, when the superstructure was rehabilitated and the structural steel was blast-cleaned, metallized, and painted. And those

12 — Washington's Crossing Bridge, built in 1904 and rebuilt after the flood of 1955. Photo courtesy of the Delaware River Joint Toll Bridge Commission.

stone abutments and piers, erected in 1834, were repointed at this time.

Both states ascribe special significance to Washington's Crossing with good reason; historians feel that the crossing of the Delaware, and the resulting victory at Trenton were the turning points in the war. As early as 1914, New Jersey created the Washington Crossing Park Commission, which laid out a park and roads, and restored and marked, for their historical significance, several old homesteads and farm buildings in the area. Research was done to accurately locate where Washington and his army crossed the river, and these locations were marked on both sides with appropriate plaques. And then authorities went one step further; on January 28, 1919, the name of the village and post office at the location on the Pennsylvania riverbank was changed from Taylorsville to Washington's Crossing. Everyone seemed impressed except the few surviving members of the Benjamin Taylor family.

FIVE

The Yardley-Wilburtha Bridge, 1835

Going downriver on the Delaware, after passing into Mercer
County at Washington's Crossing, the next stop in the old days
was called Yardleyville at Bucks County in Pennsylvania, and
Greensburg, across the river in New Jersey. After the arrival of the
modern period of simpler spellings and more sophisticated village
names, this crossing became known as Yardley's Ferry and, later,
as the Yardley-Wilburtha Bridge.

The initial ferry crossing at this location was established by an
ambitious Pennsylvania lad named Thomas Yardley Jr., a nephew
of the founder of the village, William Yardley, and son of another
equally prominent Yardley, Thomas Sr. In 1722, when legal
approval of ferries became required by Pennsylvania law, the State
Assembly confirmed young Thomas Yardley's initial ownership of
the ferry. Later generations of this family owned and operated the
local grist mill as well.

On the opposite, Jersey, shore of this river, a ferryman by the
name of James Gould operated his craft in competition with Yardley. New Jersey records of 1765 indicate that Gould's Ferry was, by
this time, owned and operated by a Mr. Howell and was called
Howell's Ferry. Soon the ferry that had been Yardley's Ferry, on

the Pennsylvania side of the river, was taken over by Mr. Howell and, of course, was now named Howell's Ferry. Howell operated on both sides of the river for a number of years.

Early in the next century, plans were underway for a bridge to replace this now outgrown and overworked ferry operation. Both states approved a privately owned bridge at this location. The Yardleyville-Delaware Bridge Company was formed, money raised, and construction finally got underway. By 1835, the privately owned bridge was completed and opened for business. It was, of course, a wooden covered toll bridge, 903 feet in length. This bridge length also included a span that passed over the Trenton Water Power Canal that ran closely alongside the river at this location. Part of the company's regulation of charges on the bridge stated that persons going to, or coming from, church on Sundays would be exempt from paying this bridge's tolls. This policy was soon expanded to cover morticians as well.

The worst flood of the century in this area took place just five years after the bridge opened for business, on that unforgettable and frigid day of January 8, 1841. Three of the new bridge's six spans were swept away and had to be replaced. The January 16, 1841, issue of the *Hunterdon Gazette* had this to say: "Not a bridge is left standing between Easton and Trenton; those at Riegelsville, Centre Bridge, New Hope, Taylorsville, and Yardleyville, having all yielded to the resistless power of the flood." Ferries went back to work and life slowed down that winter.

The bridge had not been in existence long enough to have acquired extra capital for these major repairs, and the corporation had to run into debt to pay for most of them. But what had to be done, was done, and the bridge, still a wooden covered structure, was back in operation . . . and the old ferries went back into retirement.

The bridge served the area well for the next sixty years, surviving floods and fires, and operating profitably. And during this

13 — Yardley-Wilburtha Bridge and tollhouse from Pennsylvania in 1920. Photo courtesy of the Delaware River Joint Toll Bridge Commission.

period the name of Yardleyville was simplified to Yardley, and the village name of Greensburg was changed to Wilburtha; the bridge was now called the Yardley-Wilburtha Bridge. The Yardley-Wilburtha span was starting a new life.

Business grew and thrived in Yardley and across the river in Wilburtha. There were grist mills in the area, two quarries, and during the Civil War, when sugar could not be purchased from the South, local farmers began to grow sugarcane and sell vast amounts of the finished product. Even when the war was over, this business prospered. And this stretch of the river became a prominent commercial shad fishing area. With all this activity, in 1876 the New York branch of the Reading Railroad came to town and established a station here called, simply, the Yardley Station.

But bad luck returned in the early years of the twentieth century. In the worst flood in the river's recorded history, on October 10, 1903, the old wooden bridge was swept sway, a nearly total loss. Again, optimism seemed to prevail and the major repairs were carried out. This time the replacement structure was neither wooden nor covered; it was made of steel and was open to the elements. The new bridge consisted of two steel trusses with a double driveway in each direction, and it had a wide walkway on the upriver, northern side of the bridge. This new structure was well adapted to the modern age and could easily handle heavier auto and truck traffic; according to engineering experts, this stronger steel bridge would "last forever." It did, indeed, serve the area well and proudly during the next half century. And during this same period, on December 23, 1922, the Pennsylvania–New Jersey Joint Bridge Commission, as a Christmas present to itself, purchased this handsome and safe structure from its happy private owners, for $67,500. Then, after several decades of a profitable and secure life, this bridge's "forever" came to an end.

It was the flood of August 19, 1955, caused by a pair of deadly hurricanes in the area, Connie and Diane. By this time, most of the bridges on the Delaware had been converted from timber to steel and were expected to withstand floodwater; most did. One of the four bridges destroyed was the river's last wooden covered bridge, located at Columbia in Warren County, New Jersey. The other three bridges were newer steel structures: the Easton-Phillipsburg Bridge, another Warren County pride and joy; the Point Pleasant–Byram Bridge; and, finally, the Yardley-Wilburtha Bridge crossing the river near Trenton. Immediately after this deadly flood, the front pages of newspapers indicated that these bridges would be rebuilt, and promptly, as had been the case in the past. This happened with only one of the structures; the Easton-Phillipsburg Bridge, which had lost its center span, was rebuilt and serves those cities to this day. The other three victims of this

14 — Another view of Yardley-Wilburtha Bridge, 1920, destroyed forever in the flood of 1955. Photo courtesy of the Delaware River Joint Toll Bridge Commission.

flood, the twentieth century's worst, are gone forever. Only at the Columbia site has a foot span been built. The other two bridges, Point Pleasant–Byram and, yes, Yardley-Wilburtha, are gone forever. Nevertheless, the survival score of our Delaware bridges is still quite high.

And even after the Yardley-Wilburtha Bridge ceased to exist, attention was called to the area again just a decade later, on February 21, 1964. A train rushing downriver, as it passed through Wilburtha, was derailed near the bridge site, and three of its five cars tumbled into the old Trenton Delaware Falls Canal. The relief train rushing to the scene also derailed nearby. This area received a well-deserved nickname—"Wrecked Wilburtha." This took the place, sadly, of its nickname when the fine bridge was around, "Wonderful Wilburtha."

SIX

The Upper Black Eddy–Milford Bridge, 1842

This bridge is located at the river crossing on the Delaware between Upper Black Eddy in Bucks County, Pennsylvania, and Milford Borough in Hunterdon County, New Jersey, well upriver from the earlier county spans at the Stockton and Lambertville crossings. The first settler in the area, long before there was a bridge or ferry, was a miller whose name is, today, unknown. River crossers in the earliest times had a spot where they could wade the river. After a short existence the mill burned down and the area got a name, "Burnt Mills," and the mill brook feeding the river near modern Milford was called "Burnt Mills Creek."

The first citizen and mill owner whose name is recorded was Thomas Lowrey. He built a new mill where the old one had been and called it "Lowrey's Mill." The town soon became known as Lowreytown. Thomas Lowrey also built and leased out a ferry on this part of the Delaware; it was called "Lowreytown Ferry." Its stop on the Pennsylvania side of the river was at Upper Black Eddy. Early in the nineteenth century Lowreytown's name was changed again, this time to "Mill-Ford," and by 1820, the mill and its village were both called, simply, "Milford."

A bridge was usually built across the river when demand

15 — Upper Black Eddy–Milford Bridge, first opened in 1842. Photo courtesy of the Delaware River Joint Toll Bridge Commission.

exceeded what the ferry could supply. This was the case in the Milford–Upper Black Eddy area on the river. A substantial covered bridge was considered a first-rate solution to the problem. And, just incidentally, investment in a bridge would supply the bridge's private owners with sizable incomes from the profits, it was hoped.

The first bridge constructed here was the product of the Milford Delaware Bridge Company, which was established, with Pennsylvania and New Jersey state approval, in 1839. This approval, however, did not permit the privately owned bridge company to conduct the banking operations in the area; this had been done elsewhere in the county with disastrous results.

The Milford Bridge crossed the river at a narrow location and was only 681 feet long, shore to shore. It consisted of only three spans supported by two piers. This made it safer for timber rafts to pass underneath in rushing spring freshet. The bridge was completed and opened for use on January 29, 1842, at a cost of only $27,000. However, in the first five years of its operation the company spent an additional $7,200 on unexpected repairs. Fortunately, the new structure, and its owners, just barely avoided the flood disaster that occurred in this area the year before the bridge was built, in January of 1841. This flood destroyed many of the existing Delaware River bridges, including both of the Hunterdon County spans.

This new crossing brought additional business to this part of the river valley. It gave farmers and small industrialists in the area quick access to the Delaware Canal in Pennsylvania. And this increased use brought additional funds in the form of dividends to the stockholders of the Upper Black Eddy–Milford Bridge. By 1844, businesses in the now growing town of Milford included three stores, three taverns, twelve to fifteen mechanics' shops, a flour mill, and two new sawmills that made lumber trade, here, an especially important business. The town also had many non-commercial structures, including forty-five homes, two churches, and a fine school. Upper Black Eddy on the Pennsylvania side of the river directly opposite Milford was a favorite stop for timber raftsmen in the early days. By the mid-nineteenth century the bridge brought even more business. Upper Black Eddy was booming, too. It had forty houses, three hotels, and several stores and shops.

For the rest of the nineteenth century the bridge avoided floods, or at least flood damage. In 1878, its wooden trusses were showing signs of deterioration and were strengthened by the addition of iron rods and bolts, the first indication of the use of iron in bridge construction.

The first incidence of serious flood damage did not occur until the next century, in the notorious flood of October 1903, which destroyed many bridges on the river. The Upper Black Eddy–Milford

Bridge lost one of its three spans, the one closest to New Jersey, but it was replaced promptly with another of the same type. This new span was built of timber swept downriver in the flood from the nearby, badly damaged Riegelsville Bridge. This second-hand timber was, of course, purchased at a very low cost, something that appealed to the bridge's thrifty owners. And the renovated structure appeared to be in good shape when, on June 28, 1929, the New Jersey–Pennsylvania Joint Commission purchased it for the very acceptable sum of $45,000. The Joint Commission immediately freed the bridge of its tolls.

This new owner of the bridge, the Joint Commission, was soon to discover problems with the Milford span. By 1933, the Commission came to the conclusion that their bridge, now in existence for over ninety years, was seriously unfit, and decided to have it completely rebuilt. The piers and abutments were salvageable but they had to be recapped. The entire old and reused covered wooden superstructure, everything above water level, had to be replaced with

16 — Upper Black Eddy–Milford Bridge after the flood of 1903. Photo courtesy of William Lifer.

steel trusses; and both roadway approaches to the bridge had to be rebuilt. By this time, only one other wooden covered bridge remained in existence across the Delaware, the Columbia-Portland structure in Warren County. The total cost for this extensive work on the Milford Bridge was $89,970. And just two years later, in 1935, the Bridge Commission found it necessary to underpin some of the piers and abutments to stop them from settling. The cost of this major undertaking was another $35,000.

The next exciting event in the life of this bridge was the flood of 1955. The normal water level of the Delaware at this bridge site was 108 feet; the bridge's deck was at 133 feet; and the top water level during the flood on August 19, 1955, was 140 feet. The bridge's deck, then, was 7 feet under water. But the new steel structure held up well and, when the water went down, only bridge railings, sidewalks, and a section of a wing wall needed replacement. Otherwise, this bridge had withstood this terrible flood reasonably well. When the water level dropped below the deck, the bridge was back in business.

This fine old structure, now modernized, seems to have dodged Nature's worst blows and remains in use today, still serving a lovely and rural area in both New Jersey and Pennsylvania. Local folk are proud of their region and their bridge, too . . . with good reason.

SEVEN

The Uhlerstown-Frenchtown
Bridge, 1844

The location of a ferry crossing on the Delaware River almost invariably marked the location where a river bridge would some-day be built. Such was the case at Frenchtown.

The first ferry at Frenchtown was in operation as early as 1690 but little is known of it. By 1741, the ferry was called "London Ferry." Perhaps the English agency that authorized this ferry was influenced by its name. But the name changed as regularly as the ferry's ownership, and London Ferry became, in 1759, Calvin's Ferry, and during the Revolutionary War, Sherrerd's Ferry, then Erwin's Ferry, and in the early nineteenth century, Prevost's Ferry. The river crossing here was considered to be a vital one for the rebellious Americans during the War for Independence, and the new government agreed to exempt from military service the French-town ferryman, John Sherrerd, and his three employees. John Sherrerd ferried across the river British General Burgoyne's defeated army, now all captives of war, on their way to a military prison in Virginia. Sherrerd also ferried food across the river at Frenchtown to feed the hungry rebel troops in Pennsylvania.

With the coming of peace, a wealthy Flemington business-man, Thomas Lowrey, moved to this little village and purchased large tracts of land. He also erected several mills and is even given credit for founding the village called Frenchtown. He was similarly active in nearby Milford, until he died in 1809.

In the early years of the nineteenth century, Frenchtown's manufacturing and milling grew rapidly and the ferry was soon unable to keep up with it. By 1840, plans were worked out to construct a privately owned river bridge. The bridge was called the Uhlerstown-Frenchtown Bridge but its official name was the "Alexandria Delaware Bridge." The bridge corporation's first president was Hugh Capner, who was elected in 1842. Construction was begun early in 1843, and the bridge was completed and opened for travel in the early part of 1844. The bridge company was authorized to issue $30,000 worth of stock if needed to complete the work. As it turned out, the final cost of construction was only $20,000. Fortunately, this structure's creation had just avoided the river's worst catastrophe to date, the flood of January 1841.

The bridge was a typical wooden, covered structure but a long one, with six wooden spans; the bridge was over 1,000 feet in length. The substructures of the bridge, five piers and two abutments, were strengthened with masonry, and are, to this day, still standing and still supporting a bridge.

The bridge and the villages on either end of the bridge had a moderately busy life. In the late 1840s Frenchtown had a grist mill, a sawmill, two stores, three taverns, several mechanical shops, and twenty-five residences, and, according to historians of the period, Barber and Howe, "At this place is a neat bridge across the Delaware." This bridge, which was built just after the flood of 1841, experienced a minimum of high-water damage during the nineteenth century. The flood of 1862 caused some destruction to the bridge's upper structure but it was promptly repaired.

On June 29, 1878, the business section of Frenchtown, on Bridge Street, burst into flames—arson was the official explanation—and sixteen homes and twenty-one business firms were destroyed. The wooden covered bridge, the Uhlerstown-Frenchtown span at Bridge Street, was seriously threatened but miraculously escaped damage. Unfortunately, this fire brought hard times to Frenchtown.

An equally serious calamity occurred again in the flood that struck the whole valley on October 10, 1903; it tore away two spans of the bridge near the Jersey side. These two wooden spans were replaced with steel spans but the other old wooden spans remained in place. This work took almost a year to complete and cost about $10,000. In the meantime, the old ferry, a 40-foot by 10-foot craft, went back in business, especially in getting Uhlerstown, Pennsylvania, residents over to the railroad station at Frenchtown. The old bridge was finally reopened for business, now half wood and half steel, and it still was collecting a toll from each user. It served its owners well until June 28, 1929.

On this date, the bridge, the land it occupied, and even a residential

18 — Uhlerstown-Frenchtown Bridge after the flood of 1903. Photo courtesy of William Lifer.

building on the mill's property, were purchased by the Joint Bridge Commission for the handsome price of $45,000. And on this same day all toll charges were terminated. During 1931, in the depths of the Great Depression and high unemployment, the commission went to work on their bridge, hiring hundreds of grateful and eager workers. All of the bridge's piers were recapped and repointed. The entire superstructure, the old wood and newer steel, was completely removed and replaced with a brand-new all-steel, six-span work of art. Sidewalks were laid on both sides of the bridge and road approaches were completely rebuilt. The total cost of these very extensive repairs and updating was $96,410.65. And this included lots of good pay for the many men who might otherwise be unemployed. This almost-new structure was now ready to face the future.

A major test of its future was a few years in coming, but on August 19, 1955, the deadliest and most expensive flood in the recorded history of the Delaware Valley hit the region. Homes and businesses along the riverbank were destroyed, most noteworthy, possibly, the Kerr Chickenry, where 400,000 eggs in incu-

19 — Uhlerstown-Frenchtown Bridge repaired after the flood of 1903, half original wood, half new steel. Photo courtesy of Robert Longcore.

bators were destroyed. Two hundred Frenchtown citizens were evacuated from their homes along the river; some moved into the local American Legion

20 — The new Uhlerstown-Frenchtown Bridge, 2002. Photo from the author's collection.

Hall and others were taken in by relatives and friends. Most returned home the next day to begin cleaning up.

Of this Delaware bridge, the *Easton Express* reported at the time: "About four inches of water flowed over the Delaware River Bridge at Frenchtown. The bridge was battered by debris, one huge tree lodged perpendicularly under the bridge near the Pennsylvania shore, but the bridge held fast." The sturdy, renovated bridge suffered little permanent damage and lost no service time.

The life of this fine structure, since then, has been uneventful. Only routine repairs have been made on this bridge since the 1931 rebuilding, including, through 1983: in 1949, work on concrete floors and sidewalks; 1972, repairs to riprap; 1973, approaches sealed, area cleaned and painted; 1981, Pennsylvania abutment reinforced; and 1983, bridge sandblasted, cleaned, and painted. This fine structure shines like new, in all its glory.

EIGHT

The Point Pleasant–Byram Bridge, 1855

The Point Pleasant–Byram area on the Delaware has an early history. In 1748, it had grist mills operating not from the Delaware but from a stream running into the Delaware. This was another Delaware Valley area that produced flour and shipped it to our Revolutionary armies. And certainly some of these edibles went to the British as well; they paid more.

One of the earliest millers we know by name was Jacob Stover, who, in 1803 and thereabout, owned and operated on the Pennsylvania side of the river, the busy Point Pleasant Mill. He eventually owned and operated in his lifetime a total of twenty-eight profitable mills in this area of the valley. His property remained in the Stover family for 150 years. In 1932, the Ralph Stover State Park was created in his memory and is open to the public today.

This business center on the banks of the Delaware resulted in the early establishment of a ferry here, which operated longer than ferries elsewhere along the Delaware. The Point Pleasant–Byram Bridge, which put the ferries out of business, wasn't constructed and open for business until 1855 and was one of the last of the wooden bridges to be built and privately owned on the Delaware.

R. Bridge Point Pleasant Pa

21 — Point Pleasant–Byram Bridge in 1930. Photo courtesy of the Delaware River Joint Toll Bridge Commission.

The bad luck event that assaulted this structure may have discouraged other bridge-building attempts for a while.

Just seven years after it was built and opened for business, the flood of 1862 resulted in serious damage to many bridges on the Delaware, including this new Point Pleasant–Byram structure. The wooden ruins were repaired and the Point Pleasant Bridge went back into operation. It served the area well for a number of additional years and was proudly complimented as a divinely blessed structure by James P. Snell in his 1881 history of Hunterdon County. Then, just ten years after Snell's tribute, on March 29, 1892, this bridge was struck by lightning in the early evening and burst into flame. It was almost totally destroyed.

Wisely, the bridge owners decided to build their new bridge at the same location, but to use a new fireproof product, steel. With enthusiasm, the fine new structure, one of the earliest steel bridges in the United States, was opened for business only nine months later, on December 22, 1892. It was the pride and joy of local residents. Unfortunately, the worst flood in the river's recorded history, up to that time, took place only a decade later, in 1903. Some of this new steel structure was swept downriver and was irrecoverable. The bridge required an extensive and expensive repair

22 — Point Pleasant–Byram Bridge, swept away in the flood of 1955 and never replaced. Photo courtesy of the Delaware River Joint Toll Bridge Commission.

job, another calamity that, again, deprived the bridge owners of profits.

Shortly after this unfortunate repaired bridge went back into business, the state governments of Pennsylvania and New Jersey formed the Pennsylvania–New Jersey Joint Bridge Commission to purchase available privately owned bridges on the Delaware River, make them state property, and open them for business, toll-free. When, in 1919, owners of the financially stressed Point Pleasant–Byram Bridge offered their structure to the Joint Bridge Commission, they instantly struck a deal. This bridge, the newest of the Hunterdon spans and of modern, steel construction, was the Commission's first Hunterdon County choice.

So, the Point Pleasant–Byram Bridge went back to work again and for a while seemed to prosper. Its new owner, the Joint Bridge Commission, made repairs when needed and the steel structure did fine. Then, on August 18, 1955, Hurricane Diane unloaded into the Delaware River and raised the water levels higher than ever before. The result was destruction and a loss of life unknown in earlier floods.

Most of the river's bridges by this time were steel, not wood, and most managed to survive with only minor damage. However, two Delaware bridges, the Yardley-Wilburtha near Trenton and the Easton-Phillipsburg, upriver, were severely compromised. After some debate by bridge engineers and politicians, repairs were begun at the Easton-Phillipsburg Bridge and it eventually returned to operation. It is still in service to this day. The badly damaged Yardley-Wilburtha span was not salvaged, but instead was demolished, soon to be replaced by a new structure nearby, the Scudder Falls Bridge. Two other bridges were totally destroyed in this flood. One was the fine old wooden covered bridge at Columbia, in Warren County, New Jersey. The last such span still on the river, it was swept away.

In the Point Pleasant–Byram area of the river many homes were washed away and the Pennsylvania Railroad station was wrecked. A large section of the Delaware and Raritan Canal at Byram was also destroyed. And the bridge, the modernized, steel Point Pleasant–Byram structure, had only its old stone piers still in place. The steel portion of the bridge was swept away, one twisted steel section deposited in trees on the nearby Jersey shore and another section washed up on the Pennsylvania shore.

Shortly after this disaster, the press indicated that the Point Pleasant–Byram Bridge would be replaced, but work did not begin. Finally, the Joint Bridge Commission admitted that the bridge had always been a money-loser and would not be replaced. And so the short life of this bridge, after many substantial, but unsuccessful attempts by its private owners to overcome the forces of nature, was ended forever. It was a good structure that did its best . . . but that wasn't good enough.

NINE

The Lumberville–Raven Rock Bridge, 1856

Slightly upriver from the busy and growing Stockton area and its sturdy bridge, Lumberville, in Pennsylvania's Bucks County, was hilly and difficult for travel and transportation. Although from its early life this settlement had two sawmills and a quarry, it didn't grow or prosper the way other river crossings did.

The first European inhabitants here were Swedes, who arrived in the early 1700s. As early as 1770 the village had a store, which would later also serve as its post office. In the later 1700s, a Colonel Wall settled here, built two sawmills, and began shipping his product downriver. Colonel Wall is considered the founder of this community, and at the beginning the little settlement was known as "Wall's Landing." Wall died in 1804. In 1814, the new owners of the sawmills, Mr. Heed and Mr. Hartley, renamed the village Lumberville, after its principal industry. Later, William Tinsman took over the milling, and Tinsman Bros., Inc., remained in business until the modern period. Quarrying was another local activity.

The existence of an early ferry here is unknown. As early as 1830, there was talk of building a bridge to improve the commercial life of Lumberville. In 1835 the Pennsylvania legislature approved the bridge, followed by New Jersey in 1836. Soon, a privately owned bridge corporation, the Lumberville-Delaware Bridge Company, was formed and received approval for building their bridge. Then there was a delay in the construction, possibly due to lack of funds to finance the structure. Finally, in 1853, the actual construction got underway. Fortunately, because of the delay, this bridge avoided the worst flood of the century, the disaster of 1841.

Solon Chapin, of Easton, Pennsylvania, was the head engineer for this bridge, as he had been for the construction of the earlier Riegelsville and Belvidere-Riverton Bridges. His partner in this undertaking was Anthony Fry. Available sources tell us that the bridge was finally completed and opened for business in 1856. In 1857, a supplement to the original charter exempted persons going to and from church from paying a toll. Bridge owners apparently felt this would put God on their side. This bridge was a typical wooden covered bridge, and its length of 700 feet necessitated only four spans, including one that crossed over the local Pennsylvania canal, which ran next to the river; this section of the river was relatively narrow. The final cost to erect this structure was $22,000.

This crossing was not a busy one, but it served the Lumberville businesses for half a century without a problem. Floods that hit the river in the last half of the nineteenth century, the Civil War–era calamity in June of 1862, for example, had no impact on the Lumberville span. Then, the major flood of October 1903 came to pass. Serious damage was inflicted on the Lumberville structure; one of its three river spans was swept away. The nearby Stockton Centre Bridge was undamaged and, thus, serviced the Lumberville Bridge users. The one Lumberville span that was washed away was soon replaced with a steel truss and was back in business the following year. This span was now one-third steel and two-thirds old timber; it remained that way until 1947.

The bridge, its approaches, and the stone tollhouse at the Pennsylvania end were purchased by the two states on July 21,

1932, for $25,000 and, as always, became toll-free. The combined home and office of the former toll collector was converted to the Bridge Commission's local office. As it turned out, the sale price was no bargain for the Joint Commission. Ten years after the purchase, while the Joint Commission was rebuilding one of the bridge's piers, it was discovered that the entire superstructure of the bridge, except the recent steel addition, was rotted beyond repair. After a thorough inspection, the bridge was declared totally unsafe for vehicular traffic and condemned, then closed in February of 1944.

Of course, at this time World War II was underway, and the vast amount of steel to rebuild this bridge was not available for such peacetime uses. In the meantime the Joint Commission decided that this bridge was unnecessary as a river crossing for motor vehicles. Thus, after the war, in 1947, the respected bridge-building firm of John A. Roebling & Sons was hired to replace the old bridge with a modern and safe structure

23 — Modern photo of Lumberville–Raven Rock, now a footbridge only. Photo courtesy of Thomas Drake.

24 — Modern photo of Lumberville Bridge looking at Lumberville, Pa., and the Black Bass Hotel. Photo courtesy of Thomas Drake.

designed solely for pedestrian traffic; no motor vehicles allowed. The Roebling firm went to work and removed and changed the entire superstructure to that of a modern suspension bridge. The bridge's old and original substructure, the piers and abutments, which dated back to 1855, were found to be in good condition and remained unchanged. The total cost of the major rebuilding of this pedestrian span by the Roebling group, including the remodeling of both road approaches, was $75,000.

So the substructure of the original bridge now supports this pedestrian crossing. And standing close by is that fine old two-story stone residence, part of the original bridge property, which had been the office and residence of the toll collector. It was remodeled, again, in 1968 and then became quarters for the Bridge Commission's police.

The quality of the work done by Roebling was soon tested by the river's worst disaster, the flood of August 19, 1955, and survived with only minor damage. Since then, the entire bridge deck has been reconstructed and a new lighting system installed for the safety of visitors. And the beautiful bridge has been made more easily accessible to the disabled. We should be grateful that this monument to our past still stands . . . handsome and proud.

TEN

The Calhoun Street Bridge, 1861

The second river crossing at Trenton, ferry and bridge, like the first crossing, connected Trenton in New Jersey with Morrisville in Pennsylvania. The two crossings were about a mile apart: the lower one, the site of the very first Delaware bridge, was located just downriver from the Trenton Falls; the upper crossing, at Calhoun Street, was just above the Trenton Falls. Some perfectionists also note that the Calhoun Street crossing was the first real river crossing; it was above tidewater. Both these early locations are still major river crossings.

The very first ferryman of whom we know anything, as far as the upper crossing is concerned, is James Trent, whose ferry grant, given by the New Jersey Assembly, included both river crossings, one above and one below the falls. Trent's grant was replaced shortly by a grant for a river crossing below the falls only. It is unknown whether or not James Trent ever had a ferry at Calhoun Street. We do know that by 1782 the ferry crossing at Calhoun Street was owned and operated by George Beatty. It was, of course, called Beatty's Ferry. There is some indication that Mr. Beatty had a partner identified as John Burrows. And respected historian General Stryker puts Beatty at this crossing even earlier; he tells us

25 — First Calhoun Street Bridge at Trenton, built in 1861; burned down in 1882. Photo courtesy of the Mercer Museum.

that for the Delaware recrossing by Washington's army on Christmas night of 1776, George Beatty loaned Washington a ferryboat to help him get supplies across the river.

In the nineteenth century, the ferry and property came into the possession of John Rutherford, and later, just prior to the bridge construction, the ferry was named, officially, the Kirkbride-Rutherford Ferry; Mr. Rutherford had apparently taken on a partner. But to many old-timers this ferry crossing was still referred to, with pride, as "Beatty's Crossing."

Since the opening of the Lower Trenton Bridge, the construction of a second Trenton bridge was considered worth doing. The subsequent rapid growth in the Lower Bridge's business confirmed this. Then, with the addition of railroad tracks and rail traffic to the original bridge in the early 1840s, and its structural crowding and deterioration, a second Trenton bridge became an absolute necessity. In 1847, both state legislatures increased the area in their towns where the bridge could be built, to encourage a second bridge, but nothing happened. Then, in 1859, new, more aggressive bridge commissioners were appointed by both states and they went to work. Stock in the bridge company was finally sold, and a stockholders' meeting was held in Trenton's American Hotel on August 30, 1859. Here, the site of the bridge was selected, at the foot of Calhoun Street, and plans for actual construction were laid out.

This new bridge would be a big one, the longest on the river, at some 1,274 feet, and, of course, it would be of wood construction. It would consist of seven spans that would be supported by stone and masonry piers and abutments. These piers and abutments were considered to be of vital importance and their quality was high. They are still in use to this

26 — Second Calhoun Street Bridge, still in use. Tollhouse on Pennsylvania side opened in 1884. Photo courtesy of the Delaware River Joint Toll Bridge Commission.

day. The covered structure would contain two wide roadways and two sidewalks. And an exceptionally high elevation was established for this bridge, which would, in the future, save it from flood damage or destruction. The Calhoun Street structure, another privately owned toll bridge, was built for a cost of $60,000, not a high price at the time for a 1,274-foot span. It was opened for traffic on July 1, 1861, just in time for a Fourth of July celebration . . . and the Civil War. The bridge was immediately put to use transporting federal military units and supplies southward . . . and bringing the wounded and dead home again.

The war finally ended; peace returned. Bridge traffic continued to increase, as did bridge profits, with the expanding industrialization in the northeastern United States. This bridge crossing in Trenton, nicknamed "City Bridge," managed to handle its growing business well, until June 25, 1882. On this day, wooden City Bridge on Calhoun Street caught fire and was completely destroyed, except for the piers and abutments, in one of the most spectacular fires ever to occur in the area. This fire on the river bridge, caused, apparently, by a careless cigar smoker, was virtually impossible for the horse-drawn fire equipment to reach.

The privately owned Trenton City Bridge Company promptly replaced their bridge. The contract for the new bridge was given to Phoenix Bridge Company of Phoenixville, Pennsylvania, an experienced and competent firm. The new bridge was a fireproof iron-truss structure—730 tons of iron were used—yet it was supported adequately by the old but undamaged piers and abutments. The long, new bridge, constructed by a crew of eighty-three hardworking experts, was up and ready for business promptly. This bridge, which stands to this day, was opened for traffic on October 20, 1884. The local newspaper, the *Gazette,* reported on opening day that 16 two-horse vehicles, 7 one-horse vehicles, and 175 pedestrians crossed over the new bridge on its first day. It was immediately busy again.

27 — Close-up view of the current Calhoun Street Bridge in Trenton. Photo courtesy of the Delaware River Joint Toll Bridge Commission.

From this time on, the bridge has lived a relatively trouble-free existence. It has never experienced damage from river flooding, and has required nothing other than normal maintenance. And its income was increased by permitting a lightweight trolley business, the Pennsylvania and New Jersey Traction Company, to share this substantial bridge.

On November 14, 1928, this fine structure, with its private owners' approval, was sold to the Joint Commission for Eliminating Toll Bridges for a difficult-to-resist $250,000. It was immediately freed of tolls, but the toll-charging trolley company remained in operation until 1940, when it was replaced by the local bus service.

The bridge has been regularly maintained and updated through the years. The only shortcoming might be the slow replacement of the New Jersey bridge employees' quarters, erected in 1978 without sanitary facilities; they were finally replaced by a modern structure—with toilets—in 1993. The bridge has withstood high water and floods, including the disaster of 1955, without losing a day of toll-free service.

The "new" bridge, the one that was constructed after the fire of 1882, is not now considered capable of handling modern heavy-duty trucks and construction machinery. The bridge is at present posted for a 3-ton gross weight limit and a speed limit of 15 miles per hour, and it has only an 8-foot overhead clearance. In effect, only cars and pedestrians are permitted on this toll-free bridge. This old structure is ideal for drivers who can handle 15-miles-per-hour speed limits and are more capable of relaxing with the absence of roaring modern trucks and tractor-trailers. Come, join us.

PART TWO

Midriver

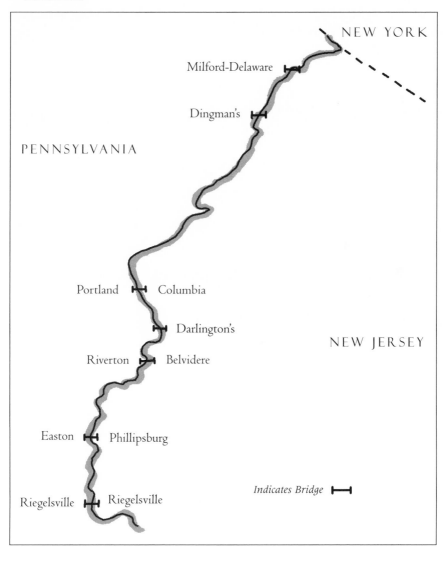

NEW YORK

Milford-Delaware

Dingman's

PENNSYLVANIA

Portland   Columbia

Darlington's

NEW JERSEY

Riverton   Belvidere

Easton   Phillipsburg

Indicates Bridge

Riegelsville   Riegelsville

ELEVEN

The Easton-Phillipsburg Bridge, 1806

The first ferry over the Delaware at Easton—before there was a Phillipsburg—operated on a charter granted by King George II in 1739 to David Martin of Trenton. It is doubtful that Martin ever actually operated the ferry himself—he leased it to others— but for years it was known as Martin's Ferry. One of Martin's early ferry operators was Nathaniel Vernon, who took over in 1755. The ambitious Vernon also kept a tavern at his ferry house at the foot of Ferry Street in Easton. Later, before the bridge was built, this same ferry was operated by, and known as, Roper's Ferry and after that, as Thomas Bullman's Ferry. It was customary to name a ferry after its current operator, and the name of the ferry changed as often as the operator did. Thomas Bullman was the last ferry owner before the bridge was opened in 1806.

The bridge over the river between Easton and Phillipsburg was the first span ever to be chartered for a Delaware River crossing anywhere on the river. In March of 1795, the Delaware Bridge Company received a charter from both states to sell stock to the amount of $50,000 to build the structure; work got underway the next year. By 1798, the bridge's abutments—foundations on each riverbank—and the piers—in the channel of the river itself—were in place. Historian James Snell tells us that at about this time a flood swept away the bridge, but there is some doubt that this happened. There were no heavy floods during this time, and other historians have concluded that high water simply carried away scaffolding left in place around the piers.

With the piers and abutments in place, construction ceased; the Delaware Bridge Company had run out of money. Both legislatures granted the company permission to hold a lottery to acquire additional funds and granted the company additional time, until 1810, to finish the job. It took a while for the bridge building to start again—the lottery had to be run and a new contractor had to be hired—and during this six-year delay the village named Trent Town, downriver, established a bridge-building company that raised the money, built the bridge, and finished the job ahead of its neighbor upriver. The Trenton Bridge was completed on January 30, 1806, beating out Easton-Phillipsburg by a full nine months. So although Easton-Phillipsburg was chartered first, Trenton completed the first bridge over the Delaware River. On the other hand, the Trenton span, wider and longer than its competitor upriver, cost a hefty $180,000 as opposed to Easton-Phillipsburg's $65,000.

The New England states were far ahead of the Mid-Atlantic states in bridge construction know-how. As early as 1662 the Great Bridge had been constructed across the Charles River in Cambridge, Massachusetts; in 1761, a large span had been built over the York River in Maine; Enoch Hale had spanned the wide Connecticut River in Vermont; and master carpenter and self-taught bridge builder Timothy Palmer had constructed the first timber-truss bridge ever, over the Merrimack River in Massachusetts. Palmer then came to Philadelphia and built the Permanent Bridge over the Schuylkill River at Market Street in 1806. Palmer had finished the Permanent Bridge without a problem, but had made a slight change in design: he completely enclosed the span, making it the first covered bridge in America. The Delaware Bridge Company

28 — The long-lasting Palmer Bridge at Easton-Phillipsburg. Opened in 1806, it was dismantled and replaced in 1896. Photo from the author's collection.

hired Palmer to complete its undertaking, and the Easton-Phillipsburg Bridge became the second covered bridge in America.

There is some debate over this claim because the Trenton Bridge was finished before the Easton Bridge, but the Trenton Bridge was not completely covered—its sides were open. The reasons for covering a bridge were at least partially cosmetic; it looked nicer, but, more importantly, it was felt that by completely covering the bridge's deck and trusses, these vital parts would be protected from the elements that caused rot and other deterioration. The fact that hundreds of covered bridges built in the last century are still in operation today attests to the validity of this theory. And although the Trenton Bridge beat out the Easton-Phillipsburg Bridge by a few months to be the first across the Delaware, the Easton-Phillipsburg lasted much longer than the Trenton Bridge and survived all the floods that occurred during its lifetime.

The fact that the bridge was covered, roof and sides, added greatly to its weight, and Palmer dealt with this by creating three timber spans supported by two stone piers. This bridge and most other covered bridges over the Delaware were supplied with numerous windows on each side, for lighting and as an opening for disposing of the animal manure that collected rapidly inside. The Easton-Phillipsburg Bridge was opened for use and the collection of tolls on October 14, 1806, but was not completed in all its details for another year. This would be the only Delaware River bridge in this area for the next twenty years. And this outstanding structure was Palmer's last bridge; he retired to his Massachusetts home, where he died in 1821. Fittingly, this fine structure was named Palmer Bridge in his honor.

Palmer Bridge took ferry traffic off the river, but the bridge's two piers in the river were considered a menace by timber raftsmen. And at high water there was little headroom under the bridge, making it a tight squeeze for the timber sailors. This bridge, the current from the Lehigh River just below, and the piers of later-built railroad bridges just below the Lehigh made this a dangerous area for timbermen, and each year several rafts and crewmen were lost in this section of the river.

The bridge itself, however, was indestructible. It survived all the floods of the nineteenth century, including the ice-filled bridge-buster of January 1841, called the "Bridges Freshet," which nearly swept the river clean of the many bridges that had been built by that time. The only other bridge between Pennsylvania and New Jersey to survive was the Trenton Bridge. The Palmer Bridge continued in trouble-free service until the end of the nineteenth century.

By then, old age, the increase in traffic over the bridge, and especially the advent of electric trolley traffic between Easton and Phillipsburg made it advisable to replace this somewhat decrepit landmark. In the 1890s, the Delaware Bridge Company decided to build a new, steel structure. For this purpose it hired a young Easton man, James Madison Porter III, to draw up the plans. Porter was a civil engineering graduate of Easton's Lafayette College, class of '86, and had worked in the field until 1890, when he returned to his alma mater as an instructor. In 1908, he became director of civil engineering at Lafayette but also worked as a consulting engineer on several highway and trolley bridges and other structures. In 1894, Porter was called upon to design and supervise the construction of a bridge to replace the old Palmer Bridge. It would be named the Northampton Street Bridge but is better known today as "The Free Bridge." This bridge would be of steel, uncovered, and of cantilever design; that is, it would be anchored to each riverbank so securely that mid-river piers to support it would not be necessary.

Porter decided that the new bridge would be built right next to the old one and that during demolition of the Palmer Bridge and construction of the new Northampton Street Bridge the contractor would maintain "a safe, convenient, and prompt means of passage for the public including electric [trolley] cars." In other words, traffic over the river would continue unimpeded during demolition and construction. The company selected for this difficult undertaking was the Union Bridge Company of New York City. They were paid $112,000 for the job. The new bridge was completed and the old one demolished—without unusual incident or delay in traffic—early in 1896. The Northampton Street Bridge would become one of the most profitable toll bridges on the river.

In October of 1903, the new structure was tested. On Thursday, October 8, most of the middle eastern seaboard was lashed by heavy rainfall. On that day alone 9.5 inches of rain fell in North Jersey, and the next day was just as bad. Horses drowned in the streets, New York subways and Scranton coal mines were flooded. Water levels on the Delaware exceeded those of any previous flood. The old Port Jervis–Matamoras Bridge was swept away, as was the covered bridge over the river at Belvidere. On the wooden

29 — The second Easton-Phillipsburg Bridge, built in 1896 and still in use. Photo from the author's collection.

railroad trestle over the river at Brainards, boxcars were pushed out on the span to hold it down, but the bridge collapsed anyway. Huge pieces of these wooden bridges, the boxcars, and assorted bungalows, outhouses, and timber rafts were rushing on the fast current toward Porter's new cantilever bridge. The *Easton Express* reported on October 12, "Great crowds were at the bridge watching the water creep up." At the forefront was James Madison Porter.

Several shanties or cottages, the homes of "Italians at Martins Creek," struck the bridge first and were crushed like eggshells. Three timber rafts then smashed into the bridge but were broken apart by the impact and passed under the span, one log at a time. Pieces of the upriver runaway bridges were shattered like kindling on the steel structure. Because of the cantilever design the bridge had no piers in the river channel and destruction at this usually vulnerable area was avoided.

In addition to the upriver spans, bridges downstream at Riegelsville, Milford, Frenchtown, and Trenton were also swept away and others badly damaged; but Porter's creation survived this onslaught with flying colors. Although some minor damage occurred, it was promptly repaired. The press dubbed the bridge, accurately, the "Gibraltar of the Delaware."

After this acid test, the "Gibraltar of the Delaware" lived on and became something of a landmark on the river. The Northampton Street Bridge was purchased by the Joint Commission for the Elimination of Toll Bridges on August 3, 1921, at a price of $300,000, and was immediately freed of tolls. It then became known to everyone as "The Free Bridge." At this time the superstructure of the bridge underwent extensive repairs and renovations. It served its twin cities faithfully through their years of glory, decline, and, now, contributes to their rebirth. James Madison Porter died in his beloved Easton in June 1928, and did not witness the near destruction of his masterpiece in the deadly flood of August 18–19, 1955.

Hurricane Connie hit the North Carolina coast on August 12, 1955. Winds of over 100 miles per hour and heavy rains lashed several southern states, but an optimistic weather bureau cancelled hurricane warnings for New Jersey. The bureau did note, incidentally, that another hurricane, this one called Diane, was following closely behind Connie.

Connie passed over New Jersey from east to west, roared into Pennsylvania, and expired over Lake Erie. In two days this vixen dumped 10 inches of water in the Delaware Valley, leaving streams and rivers at flood stage and the ground saturated. Diane arrived on August 18, dumped 6 additional inches of rain in Warren and Sussex Counties and even more in the Poconos in Pennsylvania during the night and predawn hours of that unlucky Friday, August 19, 1955.

The people of Easton and Phillipsburg went to work that morning, pleased to see a bright and sunny day. Those people crossing

30 — Easton-Phillipsburg Bridge after the flood of 1955. It survived. Photo from the author's collection.

the free bridge were amazed at the height of the water in the river but assumed that the worst was over. Before this day had passed, however, the Delaware at the free bridge would rise an astounding 43 feet, a full 5 feet above the record levels set by the Pumpkin Flood of 1903. The press called it "The Flood of the Century." The press was right.

Water continued to rise at the free bridge all day on the 19th. People who crossed the bridge that morning could not recross by the end of the day. As the water level continued to rise, so did the amount of debris in the water. Bungalows, boats, oil tanks, house trailers, and trees filled the roiling waters, and by late afternoon huge pieces of the Columbia Covered Bridge began to appear. This ancient structure had survived almost a hundred years of floods. As water levels in the river reached the deck of the

Northampton Street Bridge this debris battered the center span mercilessly.

Next morning when the sun rose on the Forks of the Delaware, a gaping hole nearly a hundred feet long had been smashed through the center span. "The Gibraltar of the Delaware" had been bested. The toll bridge just upriver, newer and built higher above the water, was still intact.

Immediately after the hurricane two Bailey bridges, temporary structures erected by the army, were put in place 250 feet upriver from the damaged span. They would remain in use for about two years. In the meantime authorities were debating whether to repair the damaged bridge or replace it entirely.

While they were debating, the flood claimed the last of its many victims. Several days after the deluge, George Stanko, after some time spent in a Phillipsburg tavern, attempted to return to Easton on the damaged bridge by walking on a bundle of telephone cables strung temporarily over the gap in the bridge deck. He lost his balance, fell into the swirling waters, and drowned.

The decision was finally made to repair Porter's old bridge, and in 1956 a contract was awarded to the Bethlehem Steel Corporation to do the job for $300,000. After the work was begun, additional damage was discovered. For example, much of the structural steel had been stretched by the force of the water and would have to be replaced. About 220 tons of steel had to be removed and replaced and the deck repaved with concrete and topped with asphalt. Surprisingly, the job was finished ahead of schedule, and the bridge was opened to traffic again on October 23, 1957.

In 1990 the bridge underwent another structural rehabilitation, this time at a cost of slightly over $2 million. At this writing, twelve years later, James Madison Porter's creation stands firm and stalwart, braced for the next "Flood of the Century." One will come someday.

TWELVE

The Milford-Delaware Bridge, 1826

This bi-state connection was achieved with the construction of a bridge on the Delaware River between Milford, Pennsylvania, and Montague in Sussex County, New Jersey, in the early years of the nineteenth century. But prior to this time, starting before the Revolutionary War, and for many years thereafter, a ferry had been busy operating between the two thriving villages. The original ferry was owned by the three Wells brothers, James, Isaac, and Jesse. As a matter of fact, at one time Milford was named Wells Ferry, indicating that this was the settlement's major business. The village of Wells Ferry became Milford in 1793, but the ferry continued in business until the bridge was built in 1836. And the ferry would get busy again in later emergencies.

Milford was a bustling settlement on the very banks of the river; Sussex County's principal village, Newton, was some twenty miles from the river. Motivation for the bridge, then, seemed to come primarily from the Pennsylvania side, and the name of the private, but two-state corporation set up to construct and operate the bridge was, therefore, the Milford-Delaware Bridge Company. This was in 1825.

The bridge—wooden, of course—was constructed under the

supervision of an experienced builder and architect, Reuben Field of Wilkes-Barre, Pennsylvania. He had already supervised the construction of a bridge over the Susquehanna River at Wilkes-Barre. Salmon Wheat was in charge of the actual day-to-day construction at the Milford site. Wheat had just recently built a bridge at the upriver village of Cochecton on the Delaware . . . twice. His first bridge at Cochecton had failed almost immediately and he rebuilt it, with the help of this same engineer, Reuben Field. This second Cochecton Bridge survived for many years.

A short segment of road was needed on the Jersey side from River Road to the bridge, and this was built by Abraham Bray and Isaac Clark. That road still exists, although the bridge itself is long gone. There is some question as to the date of the construction, but the most commonly accepted date of completion is November 17, 1826. The final cost of the structure was $20,300.

John Brink Jr. was the first bridge toll collector but he was soon replaced by Joseph Probasco, who was hired for the lucrative wage of $100.00 . . . a year. Of course he was also provided with a home to live in . . . the bridge's tollhouse. And Joseph later passed this fine job over to his brother, Charles. The toll-collecting responsibility was shared by these brothers until the bridge was swept away in the flood of 1846.

The bridge was damaged in an earlier flood, the Bridges Freshet of January 1841, which destroyed a total of nine bridges spanning the river. The New Jersey end of the Milford Bridge was damaged slightly but was promptly repaired. This minor problem, however, was followed by the fatal flood of 1846, an ice freshet that occurred in March. Huge ice chunks finished off the entire Milford Bridge.

It took a long time for the Milford-Delaware Bridge Company to refinance and erect a new bridge. In the meantime, Wells Ferry, or its equivalent, went back into operation and supplied the river transportation in the Milford-Montague area. The ferry had come back into operation, briefly, after the flood of 1841, but for a much longer time after the flood of 1846; the bridge sat unrepaired and inoperable for more than ten years. That river ferry had a new, busy, and long life.

In February of 1856, William Skinner was hired by the bridge company to supervise the construction of the new bridge. O. Clark was hired as the contractor. Although little is known of this venture, a local historian, Professor William Barr, wrote: "The bridge proved to be a worthless and treacherous affair and, a few years after its construction, it fell into the river from its own weakness." More than this is not known . . . except that the old, reliable ferry went back in business, again.

A meeting about the bridge was held in February of 1869, at which John Wallace was elected president and John C. Mott was chosen as one of the job managers. These men were determined to get a proper bridge constructed. John Mott, at this February meeting, urged the building of a suspension bridge with plans supplied by the experts of the day, the John A. Roebling Company, whose reputation was established with the construction of the suspension bridge, an aqueduct, at Lackawaxen on the Delaware. The contract for the construction of the Roebling bridge at Milford was given to John Mott, who was a primary contractor, and to his subcontractor, D. B. Rhule. The price was a modest $12,300, indicating that some parts of the previous bridge were reused. This modern and reputable suspension bridge became the pride and joy of its users and neighbors. The structure served this area well, until the fateful day of March 22, 1888.

On this day the bridge was swept away by huge blocks of ice in the thawing, rushing river. An eyewitness gave the following description, which appeared in the *Milford Dispatch* on March 28: "The river was filled with ice and the surging waters would swirl it along, crashing and grinding, and occasionally upheaving heavy ice blocks on the shore or toppling others into the seething waters. On both sides of the river, walls of ice were formed, and between them flowed the water on its mad rush to the ocean. The ice rose under the bridge and forced it upward for an instant, knocked some planks loose and tearing off other small timbers. A detached piece of ice struck a pier, some of the stones were knocked out, and the bridge was visually shaken. A large body of ice soon followed, crashing against the pier; the stone work gave way at the top as the crash came. Finally, the bridge's towers tumbled over, the cables snapped asunder, and the structure went down."

The old suspension bridge was swept away, totally destroyed, but because of its years of faithful service the public insisted upon a replacement—this time, a steel structure. Meanwhile, the bridge was, again, temporarily replaced by a ferry service, now for almost a full year. The Milford paper had this to report two weeks later in its April 6 issue: "Martin Westbrook, John Detrick, and Jacob Hornbeck will re-establish the old ferry near the Milford Bridge and have men repairing the road leading to the ferry on both sides of the river. It is expected that the ferry will be in running order by the latter part of this week; a rope cable has been ordered. 'Butcher' Van Etten will be captain of the craft."

By 1890, a new steel bridge was completed, opened for business, and was serving the public. It was soon subjected to a major test, the Pumpkin Flood of 1903, and the retired ferrymen were alerted once again. This flood destroyed many bridges on the river, including the nearby Port Jervis–Matamoras Bridge, upriver. The brand-new Dingman's Bridge just downriver was untouched, but the next span downriver, the fine old covered bridge at Belvidere, was washed away.

In a *Milford Dispatch* reporter's account on October 15, 1903, of the worst flood yet of the upper Delaware, in which he describes the destruction of the other, nearby, Delaware River bridges, as well as fourteen smaller bridges on local streams, he writes of the fate of the new Milford Bridge: "The bridge stood the ordeal well, the only damage being a slight undermining of the Pennsylvania

31 — Later Milford-Montague Bridge, built in 1890 and replaced in 1954. Photo courtesy of Alicia Batko, Montague Association for Restoration of Community History.

RATES OF TOLL ONE WAY

4 WHEEL CARRIAGE, 4 HORSES	50"
4 " " 2 "	25"
WAGON, 2 HORSES. MULES OR OXEN	25"
EACH ADDITIONAL HORSE MULE OR OX	5"
CART 2 HORSES OR MULES	25"
1 HORSE WAGON OR SLEIGH	20"
MAN & HORSE	7"
HORSE MULE OR JACK	5"
COW OR OTHER CATTLE	3"
20 SHEEP OR HOGS	20"
LARGE AUTO TRUCK	40"
AUTO TRUCK WITH TRAILER	50"
7 PASSENGER CAR	50"
6 "	25"
2 "	20"

32 — Rates charged on the 1890 bridge. Photo courtesy of Alicia Batko, Montague Association for Restoration of Community History.

abutment which can easily be repaired. The water came within three feet of the floor, according to the toll-taker, R. D. Sayers. A large tree passing under the bridge loosened a few planks on the floor." R. D. Sayers' tollhouse, where he and his family lived, was flooded with a foot of water on the first floor. They evacuated it Friday but returned on Saturday after the water level had gone down; they found everything in decent condition.

Thus, this worst flood of the upper river—to that date—did not interfere with Milford's new Delaware River structure, and it continued to provide years of reliable service. The overworked ferrymen were now retired . . . permanently. In 1922, the Joint Bridge Commission bought this toll bridge from its private owners for $31,000 and converted it to a toll-free crossing, a change that brought joy to local travelers.

By the middle of the twentieth century, however, it became obvious that the rapid increase in numbers and weight of automobile and truck traffic would soon overwhelm this structure built for the horse-and-buggy age. The bridge caretaker, who no longer had to collect tolls, was given a new responsibility; it became his sole duty to direct all vehicles weighing over five tons to the Port Jervis crossing, ten miles upriver, to avoid the collapse of the now old and somewhat decrepit Milford span.

Then, in the early 1950s, a new superbridge was constructed a quarter mile downstream from the old Milford span. This toll bridge, in existence today, is a new gateway to Milford and Pike Counties. It opened for use on December 30, 1953, and is presently operated by the Delaware River Joint Bridge Commission. Its $2 million cost somewhat exceeded the price of its predecessor. The old bridge was supposed to be shut down completely, but those who used the bridge several times a day wanted the toll-free structure to stay open for local use. For a while they got their way, but eventually the old favorite was closed for good. Only its abutments remain on both banks of the river.

Barely a year and a half after the new structure went into operation, in August of 1955, the Delaware Valley was hit with its most destructive weather on record. Beginning in mid-August with Hurricane Connie, followed immediately by Hurricane Diane, the Delaware Valley was savagely pummeled. Hundreds of lives were lost as a result of the twin storms; bridges, highways, riverfront villages were destroyed or badly damaged. Four Delaware River

33 — The new bridge, built in 1954, slightly downriver from its predecessor of 1890. Photo courtesy of the Pike County Historical Society.

bridges were swept away: the Yardley and Point Pleasant Bridges in Hunterdon County and in Warren County the Phillipsburg-Easton Bridge, which was restored later, and the fine old covered bridge at Columbia-Portland. Further upriver, Delaware River bridges were somewhat damaged but not destroyed. The span at Milford withstood this catastrophic storm valiantly, but people living nearby faced property loss and death on both sides of the river. The Jersey side was lower in elevation, and many families near the river were evacuated, fifty from Montague alone.

The Montague firemen, and a local doctor, had a problem trying to persuade an elderly woman to leave her threatened home on Mashipacong Island, just upriver from the bridge, as the water was rising on Friday morning. She refused to go with the firemen-rescuers, but when her family physician came out to the island for her, she changed her mind. They departed the flooding island in a police-paddled canoe, barely escaping. Such incidents were common during this deluge, and many had less happy results. The final death toll for the twin-hurricane storm was four hundred. This was the most catastrophic flood ever recorded on the Delaware. But the new Milford toll bridge survived and is now one of the river's major crossings.

THIRTEEN

The Belvidere-Riverton Bridge, 1836

There was certainly a ferry crossing between Belvidere, New Jersey, and Riverton, Pennsylvania, before there was a bridge, but little is known of it. Barber and Howe's history, written in 1844, states that the water power here, of the Delaware and Pequest Rivers combined, far exceeded that of any other area in the state. Ferry operation here, thus, was hard and dangerous. Small wonder, then, that on May 17, 1825, the Board of Chosen Freeholders permitted the Belvidere ferry to charge higher rates—four horses and a wagon, 50 cents; two horses and a wagon, 20 cents; foot passenger, 4 cents—and as early as 1802 Dr. Gwinnup of "Mercer," later named Belvidere, charged his patients ferriage here if he had to cross the river to see them. The ferryman, Thomas Muclehana (Gwinnup's spelling) was a patient of his, and he let the doctor ride free as payment on his bill. But more of the ferry's history we don't know except that by 1832 the crossing was busy and dangerous enough to require a bridge.

In that year, the Belvidere Delaware Bridge Company was incorporated by both New Jersey and Pennsylvania with a capital stock limit set at $20,000 for the construction of the bridge. The designer and contractor for the job was Solon Chapin of Easton,

who was also the builder of the Riegelsville Bridge at about the same time.

The early bridge-building efforts of
the corporation were disastrous. Work

was begun in 1834 on a wooden covered bridge with a length of
654 feet. The bridge was completed and opened for traffic in the
early spring of 1836. Within days, on April 9th to be exact, a heavy
spring freshet carried away two of the bridge's three spans, and
work had to begin over again. The same plans were used on the
second attempt, and the bridge was completed and opened in
the fall of 1839. A little over a year later the great Bridges Freshet
of January 1841 slashed through the valley, and again did serious damage. The new bridge survived; however, many bridges
over the Delaware were not so fortunate. It is not surprising that
Snell, in his history of Warren County, stated that due to the
expense of this almost constant repair work, as of that date (1881)

the Belvidere Delaware Bridge Company had never paid a dividend to its stockholders.

But the bridge lasted a while and became a comfortable fixture in the communities that it joined together. And soon, it even began to pay dividends, and kept up payments into the next century. Near the Pennsylvania shore in the shadow of the bridge, there was a sandy beach, and a shaded island, enjoyed by local people during the hot summer months. It became a meeting place for mothers with little children. This activity was interrupted only at bridge-cleaning time, when workers shoveled manure that had accumulated on the floor of the covered bridge and threw it out the span's windows into the river below. When possible, this task was saved for the fall or spring months when high water could be counted on to flush the stuff downriver and when bathers were not yet in evidence.

In October of 1903, the river valley got a major scouring in the so-called "Pumpkin Flood." This monster deluge hit both Riverton and Belvidere—and the bridge between—especially hard with some loss of life. Saturday, October 17, was doomsday for the old bridge. The water rose steadily and when it began sweeping over the floor, toll collector Hutchinson closed the bridge gate saying he "wouldn't give a nickel for it." The October 23 edition of the *Belvidere Apollo* reported what happened next: "At 6:40, when darkness had settled, there came a crash that could be heard for blocks. The next moment the old structure tottered for a moment, and then tumbled, to rise no more." All that could be seen at sunrise next morning were the three masonry piers poking up through the still-swirling waters. Remnants of the old bridge were carried by the floodwaters down to Easton where they were smashed to kindling against the side of the new, steel Northampton Street Bridge built there a few years before. In Belvidere, a ferry to Riverton was put into operation almost immediately.

The survival of the substructure of the old bridge was significant, for when the bridge corporation finally decided to rebuild, it set a completion date of August 17, 1904, the opening day of the Farmers' Picnic held in Belvidere. The contractor, the New Jersey Bridge Company of Manasquan, New Jersey, felt that without the necessity of building new piers and abutments they could meet this demanding deadline. The three piers required minor repairs and would be raised, hopefully, above any future floodwaters. When, on June 3, this work was started from a barge in the river under a hard-driving foreman named McGreevey, the target date was a little over two months distant. While the pier work was getting done, the steel, all 355 tons of it, was being cut and finished at the mill. Enough of the work was completed so that on the first day of the Farmers' Picnic, revelers from Pennsylvania were able to cross the nearly completed bridge. The Picnic, predecessor to the Warren County Fair, was a great success, attendance estimated at 15,000, many of whom crossed on the new bridge.

Although the span was usable, there was still work to be done on it; the contractor promised it would be completed and ready for the official takeover by the Belvidere Delaware Bridge Corporation on Labor Day, September 5, 1904. Appropriate activities for this day were planned by a local fraternal group called the Red Men, the most important of which would be the official takeover of the completed bridge. The holiday dawned bright and clear. When the chief executive of the New Jersey Bridge Company, W. H. Keepers, arrived for a final inspection, the bridge was found to be unacceptable. Bolts were missing, painting was not completed, and generally the bridge did not meet the terms of the contract. The takeover ceremonies were cancelled and many disappointed visitors—and politicians with speeches prepared—went home. The Red Men held festivities but they were but lightly attended.

35 — Murphy Jones jumps from the top center of the new Belvidere Bridge during Labor Day festivities of 1904. Photo from the author's collection.

The high point of the day was provided by Murphy Jones of Belvidere. "Murph" was a black man and local garbage collector whose contribution to the day's events would be to dive off the highest portion of the bridge into the Delaware, a 65-foot plunge. Before a large crowd, some in boats in the river, Jones climbed to the top of the new bridge and sat there, keeping the mob in a state of excited agitation, while friends of his passed the hat. When enough money was collected, Jones made his dive and the crowd roared its approval. It is said Jones put a munificent $15 in his pocket that day . . . but that this entered his stomach in liquid form. The dive of Murphy Jones became an annual event in the county seat for many years to come. The water level in the river was usually low at this time of the year and the rabid audience expected one day to see the diver break his neck on the river bottom, an incident that never occurred. The canny Murphy knew that the current from the Pequest River, which entered the Delaware here, had scoured out a deep hole in the river under the bridge.

Work on the unfinished structure crawled slowly to conclusion. People were permitted to cross on the new bridge, but because the bridge corporation had not taken over, no tolls were charged. The local paper, the *Apollo*, editorialized that the bridge should "remain free of tolls," but it was not yet to be. It was almost Christmas before the bridge was taken over by the corporation, a toll taker was hired, and life got back to normal in Belvidere-Riverton.

In October 1928, the Joint Bridge Commission brought the bridge from the Belvidere Delaware Corporation for $60,000 and on June 14, 1929, the Joint Commission gave the citizens what they had asked for a quarter of a century earlier—a free ride. Tolls were abolished. Almost

36 — Belvidere-Riverton Bridge today. "A beautiful bridge in beautiful country." Photo from the author's collection.

immediately the commission began extensive repairs to their newest acquisition, including new beams, new flooring, and major repairs to one of the piers. The work took four months and had commuters grumbling.

The supreme trial to which all bridges on the river were subjected, the Hurricane Diane flood of August 19, 1955, came and went. Although minor damage resulted, traffic over the span was halted only a day or two. Floodwaters at their peak had just barely reached the bridge floor. Designers had learned at last that the best way to make a bridge flood-proof was to build the structure above any possible high water mark. The Belvidere-Riverton Bridge serves travelers adequately to this day (2002) as long as weight and speed limit restrictions are obeyed. The river crossing is still beautiful . . . and now it is also free.

FOURTEEN

The Dingman's Bridge, 1836

America's oldest superhighway, so our traditional history tells us, was the Old Mine Road built by the Dutch almost 350 years ago to connect Esopus (Kingston, N.Y.) on the Hudson River with their copper mines in Pahaquarry on the Delaware River. The careful Dutch chose to carry their valuable ore overland, out to the Hudson River rather than risk it on the unknown and possibly hazardous Delaware River close at hand. From Esopus the ore was shipped on the placid Hudson to their New World capital on Manhattan Island. This mining venture came to naught, but the road, at least, survived; it is still there at present, and used every day.

Years later, in 1735 to be exact, another Dutchman, Andrew Dingman, began operating a ferry across the Delaware at the lower end of Old Mine Road. Migrants from the New York Colony could follow this highway on the Jersey side of the river to Dingman's Ferry and take this craft across the "treacherous stream" to Pennsylvania. Here, the traveler looking for land could continue on into the heartland of the infamous Walking Purchase, tens of thousands of acres of prime land stolen from the Lenape Indians by their erstwhile friends, the Penns. The Walking Purchase swindle took place in 1737, shortly after Andrew Dingman launched his first ferryboat. The timing was perfect. After the Revolution, a turnpike was constructed that went west from Dingman's to Bethany, a bustling village that served, for a while, as the county seat. Soon, added traffic came from East Jersey, where other landless migrants came through a pass in the mountains called Culver's Gap and followed what is now Route 206 almost all the way to the ferry. During the French and Indian War the ferry was busy taking the settlers back to Jersey; the vengeful Lenape Indians had long memories. Either way, the ferry did a thriving business.

The ferry's home base was on the Pennsylvania side of the river, and the settlement that took hold here was called "Dingman's Choice." The ferry served the area for over a hundred years, operated by four members of the family, three of whom were named Andrew. The other family member involved in the operation of the ferry was Daniel W. Dingman, better known as "Judge" Dingman, grandson of the original Andrew. The judge, who heard cases in the local court, was a man who liked his comfort; in warm weather he held court in his bare feet. And some of his sentences were strange; for example, he once sentenced a petty thief "to banishment off the face of the earth, to New Jersey." In William F. Henn's *The Story of River Road*, the judge described the village during his tenure as follows: "There are now on the place of my Grandfather's, called Dingman's Choice, eight dwelling houses principally owned by myself and family. There is also an academy, storehouse, blacksmith shop, a bridge across the Delaware, called 'Dingman's Choice and Delaware Bridge,' and a turnpike road called 'Dingman's Choice Turnpike Road.'" In spite of his lack of any legal training, he served his community well for a number of years. It was he who built and lived in the stone ferry house on the banks of the river. The house is still there.

The disadvantages of a ferry as compared to a bridge gradually became obvious: the ferry couldn't operate during high water; it couldn't operate when big chunks of ice were in the river; and it

was a dangerous trip when timber rafts were coming downriver riding the spring freshet. And, as the number of people riding the ferry increased, the boat sometimes couldn't handle them promptly and there were delays in crossing.

In 1836, the Dingmans built a bridge across the river here and retired the old ferry. The river span was wooden and a covered bridge. This design was introduced, thirty years earlier, by a carpenter named Timothy Palmer for a bridge he built on the Schuylkill River near Philadelphia. It was later followed by another of his bridges, at Easton-Phillipsburg. The Dingman's Bridge prospered and the village of Dingman's Choice grew. Under terms of the charter of this bridge, churchgoers, schoolchildren, and funeral processions were given free passage, a rule that remains in effect to this day. Then, just ten years after it was built, in 1846, a heavy rain hit western New Jersey, washing out dams and destroying property, including the Milford Bridge. The Dingman's Bridge might have survived but for the collapse of this Delaware span just upriver at Milford, Pennsylvania. The big covered bridge came surging down

38 — The piers of the Dingman's Bridge built in 1856; it collapsed in 1860. The ferry is back in business. Photo courtesy of the Pike County Historical Society.

on the rushing waters and tore away Mr. Dingman's covered bridge in an instant. In addition, his grandson Andrew, the toll-taker for the bridge, lost his entire flock of two hundred pigeons in this catastrophe. The ferry went back into operation.

A second bridge, this one also a covered structure, was built in 1850. In winter it was necessary to place snow inside the bridge so that sleighs could use it. Old-timers related that in a few years a "strong wind" lifted this covered bridge off its piers and dropped it into the river. Meteorological records indicate a tornado in the area on July 16, 1853, which the nearby *Sussex Register* described on July 30 as "the most disastrous in our history." This was probably the culprit. If so, after only three years' use of this bridge, the ferry again went back into service.

In 1856, the Dingmans had a third bridge constructed, yet another wooden span, by a firm identified in later legal documents as Skinner and Clark. Within a year, the Dingmans were charging this firm with using shoddy materials and careless construction methods; within four years the bridge simply collapsed into the river with no help from flood or wind. This was the last bridge the Dingmans would attempt. The ferry was put back in service again, and in 1868 the post office accepted the inevitable and changed the name of Dingman's Choice to Dingman's Ferry. The choice was made.

The ferry continued in operation for a number of years. In 1877, the property was acquired by John W. Kilsby, who had married Mary Dingman, the "boss's daughter." Kilsby, known locally as "Uncle Johnny," continued to operate the ferry, personally collecting the tolls. By this time a larger boat or "flat," as it was called, was put into use, guided by a cable stretched across the river. But this service was still considered too slow as compared to bridges that were in use elsewhere on the river. Dr. J. N. Miller, who lived across the river in Layton, New Jersey, but made frequent professional calls in Pennsylvania, didn't have the time or the patience to wait for the arrival of the ferryboat. He overcame the delay by pulling himself across the Delaware in a basket suspended from the overhead ferry cable. Most of the commuters, however, demanded something more convenient. Kilsby, rather than risk another bridge disaster, did nothing. Finally, in 1900, the unused Dingman's Bridge franchise was put up for tax sale in Newton, New Jersey. Three brothers, James, Will, and "E. A." Perkins, who owned the Horseheads Bridge Company of Horseheads, New York, bought it.

Thus, a fourth bridge, the one that's there now, was erected under the ownership of the Perkins brothers. The bridge itself was built largely from the remnants of an old iron bridge that had spanned the western branch of the Susquehanna River at Muncy, Pennsylvania, near Williamsport. The Perkins brothers installed this bridge, three of its five trusses anyway, on the same embankment used by the original Dingman's Bridges; they raised the bridge's piers six feet higher than the old ones, thus raising this bridge well above the usual floodwater levels. In the very destructive Pumpkin Flood of October 1903, rising river water forced "Uncle Johnny" Kilsby and family, living now in Judge Dingman's riverside stone house, to the second floor when water occupied the first. Several of Kilsby's outbuildings—the pigsty, corncrib, and icehouse—were washed away. But in spite of this high water level, the new bridge was even higher, and it survived, undamaged. The nearby bridge just downriver at Belvidere was a total loss.

At first, bridge activity didn't hint at its future sophistication. A collection sheet from July 1905 lists the bridge's activity as follows:

July 25

2-Horse and Wagons	4
1- " "	3
Footmen	2
Bicycles	2
Horses and Riders	3

July 26

2-Horse and Wagons	1
1-Horse and Wagons	4
Footmen	6
Bears	2

The last creatures were performing bears from New Jersey, heading for a star performance on the front lawn of the fashionable Delaware House at Dingman's.

And this river structure continues to survive. Second-hand bridges might be considered a poor choice for a crossing over the sometimes fierce Delaware River, but the Dingman's Bridge, owned and constructed by the Perkins brothers, has now celebrated its 100th birthday, plus a few more, with virtually uninterrupted service. The stout bridge has two lanes and, when the automobile came into use, could accommodate two cars at a time, one from each direction. In 1906, the state government offered to buy the bridge from the Perkins brothers, but they refused the offer as too low. In the 1920s, when both New Jersey and Pennsylvania began an organized effort to purchase all the privately owned bridges on the Delaware, only two bridges remained privately owned, the Lackawaxen Bridge, well upriver, and the Dingman's span. Today, only the Dingman's Bridge remains privately owned. Bears are still using this span but not, unfortunately, as paying customers. The company still carries the name it was christened with, "The Dingman's Choice and Delaware Bridge Company."

The bridge has always charged tolls. Some early tolls were: for a horse and wagon, 18 cents; a horse and rider, 10 cents; a bicycle and rider, 5 cents; the bears, in those days, had to pay 3 cents. For a horseless carriage—automobile—the cost was an exorbitant 40 cents. And prices have gone up a little since then. The rugged structure can still accommodate automobiles, but weight restrictions do not allow trucks, large recreational vehicles, or large pieces of construction equipment to use the bridge.

A toll-free crossing was again extended to schoolchildren, churchgoers, and funeral processions. In the 1920s, during Prohibition, a hearse-driving minister frequently crossed the bridge, toll-free, of course. Finally, in 1926, a rather aggressive toll collector insisted on inspecting the hearse's coffin. It contained only bottles of whiskey . . . and the driver wasn't really a minister.

This fine bridge has survived some serious floods in its century of existence. The Pumpkin Flood of October 1903 destroyed several bridges on the river, including the 651-foot superbridge just upriver at Port Jervis–Matamoras, but Dingman's Bridge survived. And in the catastrophe of 1955 and the Valentine's Day monster storm that hit the upriver area at Port Jervis in 1981, the sturdy old bridge at Dingman's also came through with flying colors.

In 1962, the Dingman's Bridge faced almost certain death when, as a result of the Delaware Valley floods caused by Hurricane Diane in 1955, the government and its Army Corps of Engineers decided to build a dam across the river just south of Dingman's Bridge at a place called Tocks Island. The owners of Dingman's Bridge were told by the Corps that their bridge would soon be underwater in the thirty-mile-long lake that would exist behind the dam. This project was eventually dropped—it was vigorously opposed by Delaware Valley residents and their sympathizers—but in the years of waiting for the end of the bridge, the structure was neglected and some deterioration took place. When the government dropped this project, the bridge owners had the equivalent of several years of catching up to do to

39 — The privately owned Dingman's Bridge, still in use in 2002. Photo courtesy of the author.

40 — Pennsylvania entrance to Dingman's Bridge. Photo from the author's collection.

get back to a safe operation. But it was done, quickly and gladly.

A recent president of the Dingman's Choice and Delaware Bridge Company was John Perkins, grandson of one of the founding brothers, James Perkins. Under John's supervision the bridge became busier than it had ever been before—1.5 million vehicles last year (2001) alone. The span is kept in good condition by a regular and thorough inspection with repairs, if needed, during a two-week shutdown every September. The present bridge, around now for more than a century, has not lost a single day because of flood damage at Dingman's Ferry. Personal injury to people using the bridge is virtually unknown. John Perkins could recall only one incident; it involved a young toll-taker who was struck by a bicycle that was breaking the speed limit as it entered the bridge. Nobody was hurt. The only other problem—since resolved—was the attempt by forbidden tour buses to cross this picturesque landmark. The buses were so wide that a vehicle coming the other way was doomed unless it could quickly back out.

Subsequently, John Oliphant Jr. was elected as the new president of the old Dingman's Choice and Delaware Bridge Company. A competent person with years of experience as its project manager and recently vice-president, he will lead the organization into its second century.

So the unique river crossing at Dingman's Ferry continues to function and to combine memories of an exciting past with indications of a prosperous future. The bridge is busier now than it has ever been before, and, undoubtedly, more lucrative. For its century of hard labor it deserves a little profit . . . and lots of admiration.

FIFTEEN

The Riegelsville Bridge, 1837

The lower Musconetcong Valley was a wonderful source of water power. Well before the Revolutionary War, forges utilizing this power—and also using charcoal made from the abundant woodlands nearby—were located here. The Greenwich Forge at Hughesville and the Chelsea Forge at Finesville were thriving at this time. Both got their raw material, pig iron, from the Durham Furnace across the Delaware in Bucks County, Pennsylvania. From this early period, then, a ferry to transport pig iron over the river was essential to these industries.

Probably the first ferry went from Durham Furnace directly across the river to the Jersey shore in what is now Holland Township in Hunterdon County. Then, a farmer by the name of Wendel Shenk, with his brother, Anthony, ran a ferry from Wendel's property in what is now Riegelsville, Pennsylvania. It crossed the river to Jersey just upriver from where the Musconetcong River enters the Delaware. This ferry was large enough for a loaded wagon and a team of horses and was propelled by oars. The village here on the Pennsylvania side was called, appropriately, Shenk's Ferry, until a man by the name of Riegel bought the property. The place was then called Riegel's Ferry and, soon, Riegelsville. The ferry landing on the Jersey shore was on property owned by a farmer and miller named Hunt; Jerseyans called it Hunt's Ferry. Later it was known as Watring's Ferry and then Leidy's Ferry as the property was sold to these new owners. By 1828, Benjamin Riegel bought the Jersey landing and the ferry was then known as Riegel's Ferry on both sides of the river.

By 1835, because of the great increase in traffic across the river and the disadvantage of ferries as already noted, the legislatures in both states decided that a bridge should be built at the ferry location. In November of 1835, they approved a private corporation called the Riegelsville Delaware Bridge Corporation as the builder and owner of the bridge. It was permitted to issue and sell stock in the amount of $20,000, but the job was done for only $18,900. This bridge corporation was denied the right to operate a sideline banking business, a privilege that had been given to some earlier bridge companies, New Hope–Lambertville in 1814 was one, with dire results. The structure was wooden and a covered bridge. It was 577 feet long and had three spans or sections supported by two piers in the river and an abutment on each riverbank. The bridge had two lanes for carriages or wagons, one for eastbound and one for westbound traffic; and two pedestrian walkways. It was built by a pair of noted Easton contractors, Solon Chapin and James Madison Porter. Porter's grandson, James Madison Porter III, would later gain fame as the engineer who designed and built the Northampton Street Bridge at Easton. The Riegelsville structure was opened for business on December 15, 1837, amid appropriate ceremonies and festivities.

Floods were the nemesis of this bridge. Just three years and one month later, on January 8, 1841, the worst flood of the century hit the Delaware Valley with a combination of extremely high water and huge chunks of ice. Six bridges on the river were destroyed and all of the others received some damage. As for the three-span Riegelsville structure, the span on the Jersey side was demolished but the rest of the bridge escaped this fate. Repairs were made by the original contractors at a cost of $9,000, and for the brief time the span was being repaired the ferry went back into operation. The bridge's unhappy stockholders were assessed this $9,000.

Again, in June of 1862, another flood hit the Delaware. A local described it: "The surface of the water at Riegelsville was covered with lumber, logs, houses, barns, pig-stys, hay stacks, bridges, canal and other boats. . . ." Notwithstanding, the bridge at Riegelsville suffered only minor damage.

In October of 1903, it would not be so fortunate. The "Pumpkin Flood" arrived on Friday afternoon, October 10. By Saturday at noon, water was flowing over the approaches to the bridge and the bridge-tender closed it. Finally, the two spans on the Jersey side were swept away, coming to rest just below Milford in downriver Hunterdon County, New Jersey. The frugal Milford Bridge Company officials would later salvage the Riegelsville timbers and use them to repair their own badly damaged structure. The remaining span of the Riegelsville Bridge was effectively destroyed, holding on by just a single, splintered timber; it finally gave way and this last section fell into the river with a crash. It was later ascertained that the floodwaters at Riegelsville reached a height at the bridge of 33.8 feet above normal, almost 4 feet higher than the Bridges Freshet of 1841, the worst flood of the previous century. The new century was starting out with a bang, indeed. The ferry went back into operation, handled by local men, Jack Edinger and Sam Sigafoos, and traveling was made easier by the inauguration of trolley service, a month later, between Riegelsville and Easton. But Riegelsville needed a new bridge.

The importance of the bridge to the area can be gauged by the promptness with which a new one was put into operation. Almost immediately after the flood, the illustrious John A. Roebling's Sons Company, with a plant at nearby Trenton, was engaged. This firm

41 — Riegelsville's first bridge, after the flood of 1903. Photo from the author's collection.

built a steel wire-rope suspension bridge 585 feet in length. It was supported by the repaired piers of the former bridge, and raised considerably. A controversy arose over the diameter of the cable to be used. Professor James Madison Porter III of Lafayette University, designer of the steel bridge in Easton that survived the Pumpkin Flood (and grandson of the Porter who built the original Riegelsville Bridge), was asked for advice. He insisted on additional cable as a safety factor and his advice was followed. The new bridge was completed and in operation on April 18, 1904. It cost the Riegelsville Bridge Company $30,000, money raised by selling additional stock.

On January 5, 1923, the *Easton Express* headlined, "There is much jubilation in the Village." The Riegelsville Bridge had been purchased by the Joint Commission for the Elimination of Toll Bridges and was now toll-free. This free use was especially appreciated when, on October 7, 1933, the Great Canoe Marathon was run on the river. The Riegelsville span, offering good views both up and down the river, was crowded with spectators, afoot and in cars, as the canoes passed underneath on their way from Easton to Philadelphia.

The suspension bridge was tested on March 11, 1936, when a flood hit the Delaware Valley. Water levels at Riegelsville were 30 feet above normal, high but not as high as in some previous floods. Nevertheless, damage to some of the piers was observed. Pier no. 1 was almost completely demolished, and the bridge was closed for a few days while repairs were made. A ferry was improvised by John Delabar and there was little inconvenience.

The worst flood in the river's history, caused by Hurricane Diane, hit Riegelsville on August 19, 1955. This flood took out the center section of the fabled Easton Bridge, carried away the old covered bridge at Columbia, and destroyed, forever, the new steel

42 — The new Riegelsville suspension bridge, still in use today after one hundred years. Photo from the author's collection.

bridge downriver at Point Pleasant–Byram. At Riegelsville, the river rose to 5 feet above the previous record deluge of 1903, the infamous Pumpkin Flood. No damage at all was done this time; traffic flow was never interrupted. The skill of the Roebling engineers—with due credit to James Madison Porter III—and the fact that the piers had been raised several feet before the new bridge was put in place, account for this miracle. Because this was one of the few bridges operable in the lower Delaware Valley, Pennsylvania state police, the National Guard, and Civil Defense workers set up headquarters in this riverside Pennsylvania village. The existence of this usable bridge gave rescue authorities quick access to both sides of the river. The Lutheran church in town, perched on high ground, served as headquarters for the Red Cross, and as a kitchen, dining hall, and first-aid station, as well. The food was so good at this dining hall that New Jersey state troopers came across the river to eat here. Indeed, years after this worst flood in the river's history, many people on both sides of the river were grateful for the staunch Riegelsville span. Floods in subsequent years have never come close to threatening this fine old suspension bridge, and it serves proudly to this day, into the twenty-first century.

SIXTEEN

The Columbia-Portland Bridge, 1869

The most recent location for a bridge over the Delaware—excluding those sterile, characterless spans built as adjuncts to the interstates—was from Columbia, in New Jersey, over to Portland on the Pennsylvania shore. It was a long wooden structure, the last covered bridge built on the river. It still thrills the memory of those who were living during its existence (this writer is among those so fortunate). This bridge was completed in 1869. Prior to that date, ferries plied the waters between Columbia and Portland. Ferry rights were first given to someone named Smith at the time of the Revolution, possibly earlier. "Smith" was a prominent name in the ferry business at one time. More prominent here was Henry Dill, who may have started out as Smith's ferryman but apparently became owner of the ferry, which was then called Dill's Ferry. Henry Dill not only owned the ferry but also a tavern at the landing site in Pennsylvania, just upriver from the present footbridge. The village was known in those days as Dill's Ferry. The village on the Jersey side that is now Columbia was, at that time, called Kirkbrides, and the ferry became known in Jersey as Goodwin's Ferry. Ferries usually changed names with each change in ownership.

Around 1800 a man by the name of William Able became owner of the ferry. He sold it to Francis Meyerhoof when the latter arrived in the village in 1812 or 1813. Meyerhoof established a glass works in Columbia, and it was his company that, in 1817, received a charter to construct and operate a bridge across the river. Then a depression hit and the bridge idea was abandoned before anything was built. Meyerhoof's enterprise, known as the Columbia Glass Works, which had promised much prosperity for the village, also failed. His company was disposed of at a sheriff's sale and never revived. His bridge charter was revoked.

Then, in 1839, a new charter was granted by both states, this one to the Columbia Delaware Bridge Company. Work began almost immediately. The structure's three masonry piers and two abutments were completed and then the work stopped, the company having run out of money, a not uncommon occurrence. For thirty years these unused piers stood there, taunting the good people of Columbia/Portland and reminding them of the shortfall. Through all this bridge turmoil the ferry continued to operate.

Then, in 1868, work on the bridge started anew. With the infusion of additional capital, the covered bridge was completed the following year. The contracting firm, the William Kellogg Company, employed many local men, S. W. Beam among them. The installation of a slate roof on the span was subcontracted to Peter Morey and Franklin Hagerman; John Henning, another local man, did most of the work on the roof. The slate came from a quarry in Bangor, Pennsylvania. The bridge was 775 feet in length and was supported by huge wooden arches whose ends rested on the abutments or piers. This was called the Burr arch-truss design and was used in the Belvidere-Riverton covered bridge also. There were two traffic lanes on the deck but no walkways.

When it was finally completed in 1869, the Columbia-Portland Bridge was the showpiece of the Delaware Valley. It was the newest of spans crossing the river and therefore boasted the most

advanced engineering techniques. It was located nearest to the fabled Delaware Water Gap and, as the gateway to this geological wonder, shared in its

43 — The Columbia-Portland covered bridge, built in 1869 and swept away in 1955. Photo from author's collection.

glory. And it was big; at 775 feet it was 130 feet longer than the Belvidere span, 225 feet longer than Easton's Palmer Bridge, and 240 feet longer than the neighboring bridges at Dingman's Ferry and Milford, upstream.

The bridge was built high enough above anticipated flood levels (the water level during the flood of 1841 being the highest recorded to date) to avoid interference with timber raft traffic as well, even in spring freshet. When the Pumpkin Flood of October 1903 hit the valley, producing floodwaters 30 feet above normal and 5 feet deep on downtown Portland streets, water lapped at the bridge's deck but caused almost no damage. On October 11, the day after the deluge, when the *Easton Express* headlined, "Only Two Bridges Left on Delaware from Headwaters to Trenton," one of these bridges was the old Centre Bridge at Stockton in Hunterdon County; the other, the Columbia-Portland span. Some experts declared this the worst flood on the river since the horrendous one in 1692.

Otherwise, the life of this bridge in the early years of the twentieth century was relatively benign and uneventful. Freshets came and went without impact, and the bridge entered the modern era with its wooden sides painted with gaudy advertising slogans. Children of the horse-and-wagon era remember the ominous rumble when a wagon passed through the wooden tunnel, and this writer can report from firsthand experience that it was equally exciting as a child to be driving in the family car through the long rattling passage with the added thrill of meeting another vehicle coming in the opposite direction. It was always a little fearful for a kid, but we still wanted to cross on this bridge.

Early in the century, Charles Newbaker came to work at the crossing as a toll collector. In those days the toll was 20 cents for a horse and wagon, 25 cents for a car, and 3 cents for a cow or sheep. On June 13, 1927, the Joint Commission for the Elimination of Toll Bridges purchased this span from its private owners for $50,000 and the bridge became toll-free. Newbaker feared for his job but was kept on by the commission as a special officer.

Shortly after the Joint Commission took over, on April Fool's day of 1929, a tornado hit the area and was especially damaging to this last remaining wooden bridge on the river. Much of the roofing and some of the siding were ripped off. It was immediately repaired, and at the same time the bridge got a complete overhaul by the commission. In addition to new roofing, steel plates were installed as reinforcement, new decking was laid, and all of the exterior timber covering on the bridge was replaced. By this time Newbaker lived with his family in a home owned by the Commission next to the bridge at river's edge. They survived the tornado nicely, apparently.

There were other exciting experiences for Bridge Officer Newbaker; his life was also enlivened, occasionally, by a heavier than usual freshet or ice flow. And in 1942, during an especially high freshet, a huge fuel tank was sent floating downriver from Port

Jervis. As retold in the *Easton Express*, May 5, 1952, "Everyone at the Portland Bridge held his breath as the large tank bobbed toward the old wooden span.

44 — The wooden bridge after a twister in 1929. It survived until 1955. Photo courtesy of the Delaware River Joint Toll Bridge Commission.

When it was about to hit the span, it made a graceful dive under the bridge and came up on the other side." The tank was not seen farther downriver. It was surmised that it got stuck in an eddy and was later salvaged, probably by a local junk man.

The old span got older and automobile traffic increased after World War II, but the bridge continued to do its job. On the Labor Day weekend of 1951 some 5,040 motor vehicles used this aged span to cross the river. Nevertheless, the Joint Delaware River Toll Bridge Commission decided to construct a new bridge over the river just below the old span. On December 1, 1953, the new structure was completed and all vehicular traffic was routed over it. The older bridge, by now the last covered bridge over the river, remained opened for pedestrian traffic only. Still, Charlie Newbaker remained on the job.

Entrance, Covered Bridge
Columbia, N.J. 477.

45 — The beautiful Columbia-
Portland Bridge just before its
destruction. Photo from author's
collection.

After fifty years of service, Newbaker, still on duty at the Columbia-
Portland covered bridge, was proud of
its longevity—and his own, no doubt.
Then on an August day in 1955 it came to an end. Hurricane Connie had passed over a few days before, leaving a deposit of rain;
and on August 18, Hurricane Diane followed close behind. The
storm had been over land several days and not much was expected
of it. However, by Thursday night six inches of rain had fallen in the
Portland area. By Friday morning, the river had risen considerably, but the sun was shining, Diane was gone, and Charlie Newbaker's bridge was still intact.

Unknown to Newbaker and to others living along the
Delaware River, the Pocono Mountain area of Pennsylvania had
been hit with ten inches of rainfall during the night. Stroudsburg,
East Stroudsburg, and lesser villages had been devastated and suffered great loss of life. This deluge in the Pocono watershed was
pouring into the Delaware River just upstream from Portland.
When Newbaker arose on this fateful morning, the river was in

his back yard and, although the sun was bright and the day pleasant, the water level in the river was climbing, not receding. Soon, water was rushing into the front windows of the Newbaker house while the officer's family clambered out of rear windows to a waiting National Guard truck. The seventy-five-year-old Newbaker retreated to higher ground but refused evacuation with his family. He was in uniform and at attention as he watched his bridge being swept away in this, the worst flood in the history of the Delaware River.

With the receding of the floodwaters of August 1955, all that remained of the old landmark were the three piers and two abutments, protruding up through the waters like the hand of a drowning man. In 1958, the Bridge Commission constructed a pedestrian bridge here, using those 1839 piers and abutments. The rest of the span was of modern design—steel and concrete. The bridge house, the former home of the Newbaker family, was sold off by the commission. Old-timers insist that on nights when there is a particularly heavy rainfall, Charlie Newbaker's ghost can be seen standing on the Portland shore, in uniform, looking at the waters rising around the old piers.

SEVENTEEN

Darlington's Bridge, 1915

This bridge got its start as a railroad bridge. It was built back in 1855 and was the first bridge of the Delaware, Lackawanna and Western Railroad to cross the Delaware River into New Jersey. The crossing was seven miles below the Delaware Water Gap at a village that would soon be known, appropriately, as Delaware Station. The original span was wood, but the railroad replaced it with an iron structure, 740 feet long, having two sets of tracks. This bridge survived several floods, including the 1903 deluge— the railroad ran loaded coal cars and boxcars out on the Delaware trestle to hold it down—and it did the job. This structure served the DL&W well, until engines and freight cars became larger and heavier. The company then decided on a larger and stronger bridge.

In 1914 the railroad built a fine new span next to the old one so it wouldn't have to move much track. When the new bridge was completed, the DL&W put the old bridge up for sale. The demand for second-hand railroad bridges being somewhat limited, the DL&W jumped at the $5,000 offer they received on December 19, 1914, from Henry V. B. Darlington of Delaware Station. The railroad executives didn't wonder at the fact that Darlington was a

46 — Entrance to Darlington's converted railroad bridge over the Delaware at the village of Delaware, New Jersey. Photo courtesy of Ruth Hutchinson Gommoll.

doctor of divinity; that indeed, he was an Episcopal minister who preached in that fine old church at Delaware Station, as well as those in Hope and Belvidere; nor did they trouble themselves about what (in God's name) the minister was going to do with an old iron railroad bridge. His money was as good as anybody's; they made a deal.

It wasn't long before the good doctor of divinity's intentions were known. He converted the railroad bridge to a vehicular bridge, built roadways to it on both sides of the river, and constructed a little house for toll collections and living quarters at the Jersey end of the span. This humble man of the cloth had sound entrepreneurial instincts, for he realized that the auto was just coming into its own and that New Jersey and New York motorists would be anxious to explore such natural wonders as the Water Gap, the Poconos, and some of the exotic spas on the far shore. Almost from the beginning the crossing was busy with

auto traffic. Henry Darlington decided to incorporate his enterprise with the impressive name of the Knowlton Turnpike and
Bridge Company. But from the beginning, this crossing was
known, simply, as "Darlington's Bridge."

When it went into operation, Darlington's Bridge had only
one serious competitor, Myers' Ferry. For generations this ferry
had been carrying vacationers, and others, across the river to Pennsylvania to Myers' own hotel or to other resorts, such as Tuscarora
and the Water Gap. Owners of the ferry operation after the Myers
family sold it were Charles Hartzell and, finally, Joseph F. Klein,
local hotel and ferry owner. After a ferry accident that killed four
passengers—Klein's ferry operator, Edward McCracken, was on a
dinner break, and Klein himself was operating the ferry—Klein
was blamed for the accident and the deaths of three women and a
son. He promptly put his ferry up for sale and, just as promptly,
fired the operator, Ed McCracken. When Darlington became the
owner of the railroad bridge and converted it to a pedestrian and
auto span, he also picked up the ferry for a song. He immediately
shut down this competitor and hired the former operator, Ed
McCracken, and his wife, to become the toll collectors—and tollhouse residents—on the bridge.

The only other bridge in the area easily accessible for auto traffic was the covered bridge at Columbia, upriver. Fortunately for the
minister, the traffic from North Jersey and New York City coming
west on Route 6 (now Route 46) came to Darlington's Bridge first
and most drivers used it. Old-timers recall that the McCrackens
collected the tolls in bushel baskets and remember seeing these
good-sized containers filled to the brim with quarters and fifty-
cent pieces. A myth existed that Mrs. McCracken at one time took
a washbasin full of silver coins to the middle of the bridge and
dropped them, by mistake or otherwise, into the water. For years,
local youths favored that part of the river for swimming and div-
ing, but nobody ever reported finding anything of value.

47 — Darlington's Bridge on the left, busy as usual. Photo from author's collection.

In spite of the presence of large sums of cash, the tollhouse was never robbed. This was credited to the innate honesty of people in those days, to the bridge's ownership by a "man of the cloth," and to the fact that the toll collectors kept two vicious Airedales, Duke and Totsey, at the tollbooth, day and night. So effective were the dogs that never was so much as a quarter missing from the proceeds. Indeed, if motorists inadvertently dropped coins on the ground here, they rarely attempted to exit their cars to retrieve them. Unfortunately, in spite of the loneliness of their job, especially during the quiet evening hours, visitors rarely paid a call on the McCrackens, thanks to Duke and Totsey.

Darlington's Bridge prospered until 1932, when the Joint Commission for the Elimination of Toll Bridges, a Pennsylvania–New Jersey government agency, purchased it from the minister, after lengthy negotiations, for $275,000. This high price is a tribute to Henry Darlington's bargaining skills; the Joint Commission paid only $60,000 for the much newer steel bridge just downriver at Belvidere–Riverton and just $50,000 for the nearby covered bridge at Columbia–Portland. Now, in 2003, only one Delaware toll bridge remains in private hands, the span owned by the Dingman's Choice and Delaware Bridge Company at Dingman's Ferry, Pennsylvania.

Knowlton Township residents, as well as city folk bound for Pocono Mountain resorts, rejoiced at the abolition of tolls over the river here. A whole generation grew up thinking that crossing the Delaware River would be forever free. Then in 1953 a modern toll bridge was erected at Columbia by the Delaware River Joint Toll Bridge Commission at a cost of over $4 million, a far cry from Henry Darlington's five-thousand-dollar span. Ironically, the toll remained the same—a quarter. The good pastor, who lived long enough to see all this transpire, must have shaken his head in disbelief.

Sadly, the Joint Toll Bridge Commission decided to dismantle the old Darlington Bridge—maybe it was an embarrassment to the Commission—and in spite of strenuous legal action to prevent the loss of this historic—and free—river crossing, Darlington's Bridge ceased operations on April 3, 1954, and was immediately demolished. That other free bridge, the covered span at Columbia-Portland, was washed away the following year in the flood of 1955 and was never rebuilt. Now the closest free bridge for the people of the area is at Belvidere, but for us cheapskates, it's worth the trip.

PART THREE

Upriver

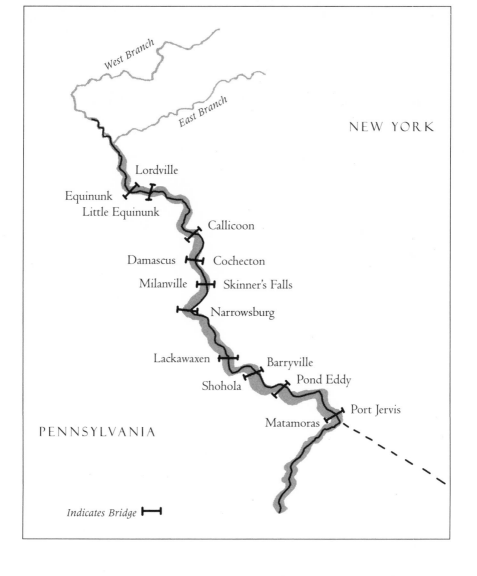

West Branch

East Branch

NEW YORK

Lordville

Equinunk
Little Equinunk

Callicoon

Damascus   Cochecton

Milanville   Skinner's Falls

Narrowsburg

Lackawaxen   Barryville
                        Pond Eddy
          Shohola
                              Port Jervis
              Matamoras

PENNSYLVANIA

Indicates Bridge ⊢—⊣

EIGHTEEN

The Narrowsburg Bridge, 1811

"This place derives its name from the fact that the Delaware is here compressed by two points of rock into a deep, narrow channel." So we are told in an 1872 edition of *French's State Gazetteer*.

"Narrowsburg" was, then, an appropriate name for this location; its other name, "Big Eddy," was less accurately descriptive. Narrowsburg was situated on the eastern side of the Delaware River in New York State. The Erie Railroad crossed the river here, laid track, and established the local railroad station.

Another active use to which the river was put during the nineteenth century was rafting of locally produced timber down with the current. These timber rafts, a thousand a year during its prime period, passed through the extremely slender channel at Narrowsburg, mostly in the spring when the water was high and powerful and the river crowded with rafts. It was a dangerous spot for timbermen but after the bridge was built, a safer spot to cross the river.

Although suspension bridges were a popular choice for spanning the upper Delaware River, a few bridges preceded the coming of this engineering marvel and were given a less ingenious design. Such a structure was the first Narrowsburg span, which was the

first bridge of any kind built across the upper Delaware between New York and Pennsylvania. This structure, built in 1811, was a wooden covered bridge.

This span was built by the Narrowsburg Bridge Company, shortly after it received its charter in 1810; it was completed in 1811. The wooden structure was only 184 feet long but had an unusual width of 25 feet. And it passed over the deepest spot in the river north of faraway Trenton; the river was 113 feet deep here. The bridge's construction was paid for by private owners who had purchased stock in the bridge corporation. But, of course, these owners would get their money back, and more, by charging sizable bridge tolls. Early charges were: a four-horse carriage, $1.00; a two-horse carriage, 75 cents; a one-horse carriage, 37 cents; foot passengers and cattle were 6 cents each. Historian Arthur Meyers, in *Crossing the Delaware River via Toll Bridges,* remarked, "Considering the value of money at that day, these rates were certainly high enough to suit the most avaricious stock holders." A contemporary referred to it as "a monstrous bridge."

This span, like many such structures at this early period, had a short life. It was destroyed by a high ice flood in 1832. A new bridge was built here that same year, somewhat higher than its predecessor. Nevertheless, it was destroyed by a flood in 1846. Its replacement was constructed two years later, in 1848. This span, too, was a wooden covered bridge, but it was 250 feet long and thus had more land support and greater stability than the previous bridge; the new structure was 22 feet wide and had the increased height of its predecessor. The bridge's owners were optimistic about this much longer structure and waited anxiously for its first test. This finally occurred on February 8, 1857. In the Narrowsburg area, a flood on the ice-filled river carried away part of the Erie Railroad bridge over the Delaware and also destroyed a hundred feet of additional Erie track at nearby Cochecton. Worse yet, the almost new bridge just nine miles upriver at Cochecton was totally swept away. But the new and improved Narrowsburg structure survived, nicely. Local folk were proud, and especially so were the bridge's owners, engineers, and builders. They had by now, apparently, learned some lessons; this wooden structure lasted almost to the twentieth century—1899, to be exact.

This bridge, indeed, lasted a long time, but like most wooden structures, eventually died, this one of old age. The new replacement bridge, under construction in 1899, was made of iron, by this time a more common bridge-building material than wood. The contracting firm that erected this span was the reputable Oswego Bridge Company; plates with its name and the date were mounted on either end of the bridge and proudly photographed for posterity.

But even the new iron bridge didn't stay new forever. Eventually, it needed some updating and in the 1920s, $25,000 was spent strengthening the iron structure to enable it to handle the heavier, modern traffic that was beginning to appear. This iron bridge, built at the turn of the century, was taken under consideration, early, as a New York–Pennsylvania Jointly Owned Delaware Bridge. Former assemblyman Cross of Callicoon had introduced the toll-free state ownership idea early in the century and kept up this activity even after he retired from state politics. He finally was successful. On January 12, 1927, the *Narrowsburg News-Times* announced, "Narrowsburg Bridge free Friday afternoon at 1:30 p.m. No more tolls will be collected."

This fine structure, now state-owned, served the area well for half a century. One of the chief products hauled across the span was coal, and it was transported tons at a time. Other commercial products were also trucked to the other side. After a half century of this hard labor

49 — Bridge and tollhouse built at Narrowsburg in 1899. Photo courtesy of the Tusten Historical Society.

a new replacement span was needed, and this structure was erected in 1952–53 by the two-state bridge commission for

50 — The newest bridge at Narrowsburg, opened in 1953. The old bridge still stands behind it. Photo courtesy of the Tusten Historical Society.

$489,674, a good price for the period. The old bridge was left intact and in use until the new bridge was opened and operating so that bridge users at Narrowsburg would not be without a bridge for even one day.

The new bridge was dedicated on August 31, 1953, at a ceremony attended by top officials from both New York and Pennsylvania; a formal dinner was held at the Peggy Runway Lodge. This new span was erected shortly after the end of World War II, just in time to handle the greatly increased postwar traffic. Two superhighways leading to the bridge, the Wayne County Turnpike and the Mount Hope–Lumberland Turnpike, were completed, connected, and soon opened for business; these busy highways added greatly to the bridge's activity.

The bridge today is a handsome, modern crossing, though a short one. As a modern steel and concrete creation it seems that it might last forever. The good people of beautiful Narrowsburg hope so.

NINETEEN

The Cochecton-Damascus Bridge, 1819

This bridge is located at Cochecton on the upper Delaware in Sullivan County, New York. The village of Damascus is its anchor on the Pennsylvania side. The first official river crossings here were conducted by a ferry, which was in operation until the bridge took its place. This span was privately built, owned, and operated by the Cochecton Bridge Company, a local corporation owned mostly by Cochecton and Damascus citizens. The first structure was an early one on the river, built in 1817 and opened for business in 1819. It was a 550-foot wooden bridge supported by only one pier in the middle of the river. Major Salmon Wheat of nearby Orange County, in New York, was the builder. The War of 1812 had sparked a great deal of local patriotism; free passage on the bridge was immediately given to all military men using this structure.

Historians tell us that this bridge, after only a very short life, collapsed into the river of its own weight; the cause was defective construction. Ferries went back to work until 1821, when a second bridge was built here. Major Wheat was again given the job, in order to make up for his previous failure, no doubt, or perhaps the architect was blamed. Wheat got some help, this time, from bridge engineer Reuben Field. The second bridge was designed by Field and Wheat to be much more durable than the first. The new structure was 600 feet in length, had three spans and two piers, and was supported by arches made from massive white pine timbers. The bridge owners had great confidence in their new toll span when it opened for business. Their confidence was justified. The bridge operated without problems for a quarter of a century and made considerable money for the owners. Salmon Wheat, now that his bridge skills had improved, went on to the construction of the Delaware Bridge at Milford, Pennsylvania, and did a fine job there. And he had the help of his friend and coworker, engineer Reuben Field, at Milford, also.

But wooden structures don't last forever. In the spring of 1846, the fine Cochecton Bridge was hit with high and potent floodwaters that destroyed the pier on the Pennsylvania side of the bridge; and when the pier fell it carried two spans with it. A ferry went back to work.

In the spring of 1847, the Cochecton Bridge Company contracted with the Thayer and Benton Company of Springfield, Massachusetts, for the rebuilding of the old bridge for $10,000. This new, or newly rebuilt, bridge was known as Benton's Bridge. Before the bridge rebuilding was completed, however, this construction firm failed financially, and the Cochecton Bridge Company took over but hired Mr. Benton to continue the work. The bridge opened again in the winter of 1847–48. Unfortunately, spring floods and high water on the Delaware at this time were especially dangerous and destructive. Timber rafts were being assembled upriver, getting ready to start their downstream voyage. These new rafts and their unassembled timbers were swept down the river like fierce battering-rams and were especially destructive to bridges and, indeed, to anything else in the way. This new bridge lasted only until the spring high water of 1848, when the span on the New York end of the bridge was smashed by floating timber and collapsed.

After this calamity, a new span and a new pier were constructed and the bridge opened again in the winter of 1849–50, but the ferry remained close by and alert. Before the year 1851 was over the span on the Pennsylvania end of the bridge again collapsed in high water and the ferry, once again, was called back to service. This time the ferry remained in operation here for almost four years, until 1854, when a new bridge was built and opened for business. This bridge, which replaced the Benton work of art, was called the Chapin Bridge; the contractor hired for the replacement was Salon Chapin of Easton, Pennsylvania. Then, on February 8, 1857, this portion of the river was hit again, this time with a major and destructive ice flood, and the entire new Chapin structure was smashed and swept away. The Erie's railroad bridge nearby was lost also. The ferry again went back into service and stayed there until a new bridge was built. Ferries were a vital aid to river crossing at Cochecton. Historical records tell us of some of the ferrymen who worked hard transporting large numbers of people across the river. Two outstanding ferrymen who were busy as bridge substitutes were Albert Smith and, later, Thomas O'Reilly; and there were more.

On February 1, 1859, another new bridge, this one a wooden covered bridge, was constructed and opened for business. Significantly, all the bridge's piers were raised, some by four and others by six feet. This bridge then functioned adequately for a number of years, and optimism began to reappear. In 1872, the middle span of this bridge's three spans settled, and repairs were required, but then the bridge promptly opened again. This structure remained in reasonably good operating condition into the twentieth century and handled a marked increase in something called tourist traffic.

As nearby city people, Manhattanites and northeastern Jerseyans, began to discover the exceptional beauty in the nearby Delaware Valley, the tourist movement into the area began to take hold. It started in the last quarter of the nineteenth century and contin-

51 — The Cochecton House, a busy hotel near the Cochecton Bridge over the Delaware. Photo courtesy of Bob and Mary Ann White.

52 — The Erie Hotel at Cochecton, N.Y., across the street from the Cochecton House. It was a busy town in the late 1800s, thanks to its bridge. Photo courtesy of Bob and Mary Ann White.

ued strongly into the twentieth. The secluded upriver areas, especially if blessed with a stop on the Erie Railroad and a nearby bridge over the river to facilitate tourist exploration, were especially attractive. Confidence returned to the good people of Cochecton with the arrival of a large and prosperous visiting population. The village of Cochecton, with the fine new bridge as its center, soon built its first luxury hotel in the same area, a three-story skyscraper, which, because of its immediate success, was followed by the construction of a second one. The Cochecton House and the Erie Hotel, as well as other inns, shops, antique stores, countryside tours, and

53 — This replaced the wooden covered bridge in 1902 and was replaced in turn by a modern steel bridge in 1951. Photo courtesy of the Tusten Historical Society.

a quaint and now prosperous covered bridge, brought tourists to the whole area but to Cochecton, especially. This was a prosperity that would go into decline only with the world-wide Depression of the 1930s.

Meanwhile, Mother Nature attempted to end the tourist boom ahead of time. In 1902, the terrible upriver flood of that year carried away from Cochecton in its rushing waters, barns, outhouses, a local church, and the local school, among other things. And the flood totally destroyed the Cochecton Bridge, the last wooden covered bridge in the area. A modern, iron structure was built immediately and greeted the tourists when they returned the next summer. Then, just two years later, in the early morning of March 26, 1904, another flood, the spring ice breakup, hit the upper Delaware Valley. In Cochecton, homes, barns, and outhouses were swept from the river's banks, and two bridge spans were lost. The damage to the bridge and other affected structures was promptly repaired, and they were again put back into operation for the tourist season. The new iron bridge held up well after this, and served its stockholders adequately until the ownership changed in 1923.

On January 13 of that year, the Joint Bridge Commission of Pennsylvania and New York offered the bridge's private owners the fair sum of $24,951 for their bridge; they accepted. The bridge became a state-owned, toll-free structure. Repairs and modernizing were promptly undertaken, and this span served the area well for another quarter of a century. Then, in 1950, the Commission decided the time had come for a replacement bridge better adapted to modern transportation vehicles. Work got under way promptly, and before long a new wide and handsome three-span structure with spacious sidewalks for local strollers was up and almost ready for use. This eight-hundred-thousand-dollar showplace was finished on November 17, 1951, and was officially and enthusiastically dedicated before a cheering crowd. An abutment from the old bridge still stands on the Damascus shore in fond memory, and, on a nearby hillside, the ancient but immaculate village cemetery overlooks the new span. This event was followed four years later by the major river disaster of August 1955, and, although downriver it was a record-breaking killer, further upriver, including Cochecton, it resulted in only a slight water level increase.

This handsome bridge now handles easily the heavy and toll-free traffic that crosses the river here. Today, after half a century of hard work, the span has captured local pride. This bridge is, indeed, "high, wide, and handsome" and promises great longevity. The good—and long-suffering—people of Cochecton deserve it.

TWENTY

The Port Jervis–Matamoras Bridge, 1852

Like most Delaware bridge sites, ferries were here first. An early ferry operated from the foot of a Port Jervis road, soon christened Ferry Street, and crossed the river to Matamoras, Pennsylvania. This ferry was owned and operated by the Westfall family. There was also another ferry in operation in this general area. It was located at a place then called Carpenter's Point but today known as Tri-States, where New York, New Jersey, and Pennsylvania come together. This ferry ran two boats from the narrow neck of land between the Delaware and Neversink Rivers; one boat went to the New Jersey shore of the river, another to Matamoras in Pennsylvania. Carpenter's Point was named after the ferry's first owner, John D. Carpenter, who ran this operation for many years and was later replaced by his equally capable son, Benjamin. These ferries had a long life as ferries go; they continued in service until the first bridge opened for business between Port Jervis and Matamoras, in 1852. And they would return to service from time to time, after that, in emergencies.

The bridge was unique for this area. It was built by the New York, Lake Erie, and Western Railroad Company as a combination railroad and wagon bridge, reminiscent of that first and short-lived

54 — This fine Barrett suspension bridge between Port Jervis and Matamoras was built in 1875 but destroyed in the 1903 flood. Photo courtesy of the Minisink Valley Historical Society.

Delaware bridge built at Lower Trenton many years earlier. This bridge, too, experienced a relatively short life. After eighteen years of heavy-duty rail traffic in addition to the horse-and-wagon service, the bridge was destroyed, in March of 1870, with the help of a hurricanelike windstorm. The railroad by this time decided to build its main rail line elsewhere, and soon constructed two double-track railroad bridges crossing the Delaware upriver from Port Jervis.

The Erie Railroad—having shortened its name—gave up its ownership of the Port Jervis Bridge, or what was left of it, to the Lamonte Mining and Railroad Company. This firm was not acceptable to the two states and a new contract was awarded, instead, to the Barrett Bridge Company of New York. In 1872, this well-established bridge corporation built the new structure joining the two towns of Port Jervis and Matamoras, but at a slightly different site on the river from that of its short-lived predecessor. This location has been occupied by bridges ever since; and they have all been referred to as the "Barrett Bridge." Fortunately, this new structure was built for horse and wagon only; it was not to be a railroad and horse-and-wagon combination. The bridge was a two-span, cable suspension toll bridge. It, too, had an extremely short life. This bridge was swept away on March 17, 1875, in an ice freshet that first carried away the sturdy river bridge of the Erie Railroad where it crossed the Delaware just four miles upriver. The Erie Bridge rode the ice downstream, where it smashed into the new Barrett Bridge and took it away also. Both of these bridges were promptly rebuilt. The Barrett Bridge Company signed a contract with the Watson Manufacturing Company on March 30, 1875, just two weeks after the flood. The newest bridge was, wisely, constructed four feet higher than its predecessor. This Barrett Bridge lived a long life.

But the bridge began to show maturity and then old age in the last years of the nineteenth century. A timber raftsman, Frank Walker of Walton, New York, remembers its condition at that time keenly. He was coming downriver with his crew on his timber raft in high water and at a great rate of speed. As he passed under the Barrett Bridge he noticed a circus caravan crossing the river on the old span toward Port Jervis. Missing from this lengthy parade was the herd of elephants for which the circus was best known. The raftsman found them soon enough. As he passed under the bridge, Frank Walker, with his crew, were suddenly in the middle of the swimming herd of beasts. Local authorities had insisted that the heavy elephants not be permitted to use the bridge; in its old age, the structure was shaky, and threatened collapse. Walker attempted to avoid the animals and was almost successful. Then, the raft hit the herd's leader, badly cutting its ear. The angry elephant attempted to climb up on the raft. It got partially aboard, but finally the terrified crewmen managed to drive the animal back into the water. They were safe but thoroughly traumatized. So were the elephants.

The bridge manager who would not allow elephants on his bridge was using good judgment. The old, somewhat flimsy structure was hit hard by the flood of October 10, 1903. Although when it was originally built, in 1875, it had been erected four feet higher than its predecessor, the old, but still handsome, structure was not high enough and was swept away on the morning of October 11, 1903. And this accident had a death toll. Five people were trapped on the bridge when it was swept away; one, Theodore Durant, managed to save himself, but the others on the bridge were killed, including Reverend Father Archangel of the Franciscan Order of Paterson, New Jersey.

The ferry went back in business again and provided almost the only transportation across the river. The *Port Jervis Evening Gazette* of October 27, that year, tells us that the ferry to Matamoras started operating just the day before, and that already it had transported the Milford Stage Coach with its team of horses. And the ferry would operate each day until midnight with A. B. Quick running things. The alternative to crossing the river by ferry was walking across on the railroad bridge or rowing a personal boat, but most preferred the ferry, which charged only two cents each way. This ferry was in operation until the new bridge was built and opened on August 1, 1904.

This bridge was erected during the cold and flood-prone winter and spring months by the Oswego Bridge Company, a reputable and competent firm. Many of their employees had worked for the Horseheads Bridge Company at the construction of the nearby Dingman's Bridge just four years earlier. Both of these bridges still stand in service.

The structure built by the Oswego Bridge Company at Port Jervis was a two-span arch truss bridge, with one span 320 feet long, the other 330 feet long. It was an all-steel structure this time, and, in addition, its abutments were two feet higher than its predecessor's had been, and the predecessor's were four feet higher than those of the bridge it replaced. This new Barrett Bridge was the pride and joy of the citizens of this area, with good reason, as it turned out; it had at last passed the test of time.

On March 23, 1922, after almost twenty years of competent service, the Barrett Bridge was taken over by the Pennsylvania–New York Joint Commission for Elimination of Toll Bridges. The newly retired toll collector, Jacob Miller, turned in the keys and threw away his coin box. The price was $153,250, and the Joint Commission then spent another $15,000 to strengthen and update the steel Barrett Bridge. And, of course, tolls were eliminated. This resulted in a real improvement in the area, thanks to the new use of the abandoned toll-house on the riverbank. The retired building was purchased in 1929 by a pair of sisters named Hoey—their first names were Flo and Jean—and they converted this

55 — This steel span replaced the Barrett suspension bridge in 1903. This bridge was replaced by the present (2002) Port Jervis Bridge in 1939. Photo courtesy of the Minisink Valley Historical Society.

structure into the now long established Flo-Jean's Restaurant, a first-rate local eatery in Port Jervis.

The Barrett Bridge, now a toll-free structure and, like most of us from the 1920s, somewhat the worse for wear, was sensibly declared, in 1939, inadequate to handle the great increase in truck and auto traffic. The R. C. Ritz Construction Company was hired to build a new double-spanned steel bridge with a 44-foot width, which made it a four-lane roadway. It was the largest New York State–Delaware Bridge and cost $380,000. And it was still toll-free. This new structure was named, officially, the Mid-Delaware Bridge. Some local folks called it "The Big Barrett Bridge."

While the new, much larger bridge was being erected, the old span was moved slightly upriver and temporarily put back into business for busy drivers, until the new structure was completed. Otherwise, local bridge traffic would have to use the next closest alternative span, the one at Milford, and it was old and decrepit and probably couldn't handle any extra traffic. So things worked out well for bridge users.

The new bridge has been exposed to its share of high water, but, because of its exceptionally sturdy and elevated construction, this span, so far, has been flood-proof. The worst flood of the twentieth century, the August 1955 calamity, did little or no harm to the new bridge. Maybe flood damage on the Delaware is a thing of the past. Let's hope so.

TWENTY-ONE

The Barryville-Shohola Bridge, 1856

There is every indication that the earliest human crossers of the
Delaware in this area were Native Americans. Their trail followed
Shohola Creek to where it enters the Delaware, and undoubtedly
they crossed the river here to a gentler trail location along the east
bank. In 1762, and for a while thereafter, large groups of settlers
from Connecticut crossed the Delaware at Shohola going westward on their way to the Wyoming Valley. As is to be expected, a
ferry followed soon after, at the end of the eighteenth century,
and, for many years, operated on the river between Shohola, in
Pennsylvania, and a settlement that later became Barryville, in
New York. The appearance of the Delaware and Hudson Canal on
the Barryville bank of the river in 1827 and the construction of
that fine railroad, the Erie, on the Shohola side of the river in 1849
brought a great increase in business to the area and made the erection of a bridge absolutely essential.

In 1854, the construction of a span connecting Shohola with
Barryville was planned by a private concern, the Barryville and
Shohola Bridge Company, of which Chauncey Thomas of
Shohola was the first president. Mr. Thomas attempted to hire
bridge expert John A. Roebling, but Roebling was busy on

another span in Niagara, New York, on the Canadian border. The Niagara bridge was to be a two-level, 821-foot structure; the lower level for horse-drawn vehicles, the upper level to handle railroad trains. It would be the first railroad suspension bridge. This undertaking required all of Roebling's personal attention and he could not, at this time, take on the construction of the Shohola Bridge. Roebling gave Thomas verbal instructions when he visited the Niagara work site, and followed this up with written instructions. Chauncey Thomas thus supervised the construction work being done by a crew of inexperienced local men he had hired. The result was a single-lane, two-span suspension bridge having a width of only 10 feet. It had a single span 495 feet in total length with no central support. The bridge was elevated 25 feet above normal water level to avoid flooding. The cost of this bridge was $9,000.

A respected historian, John Willard Johnston, who knew Chauncey Thomas personally and visited the area during his ownership of the toll bridge, insisted that Thomas, as the builder, was grossly incompetent. He was, indeed, inexperienced in bridge construction. And so was Roebling, with his two-story railroad bridge. That unique span lasted only forty years and then had to be replaced by a bridge that could handle the much heavier trains that had come into use.

When the Barryville-Shohola Bridge was built in 1856, however, the Delaware and Hudson Canal on the Barryville side of the river and the Erie Railroad on the Shohola side of the river welcomed it enthusiastically. The bridge got busy almost immediately, servicing both the canal and the railroad, and their customers, as well as local tradesmen and farmers.

A common enemy in this upriver area was the windstorm. Chauncey Thomas's wire-rope span without central support was attacked by a most severe wind-with-rain calamity on July 2, 1859, its third birthday. The bridge was almost a total loss; only the piers and abutments survived. A man and a woman crossing on the span when it collapsed were seriously injured but survived. The local ferry, after only a three-year rest, was still in good condition and went back in business.

The bridge was promptly rebuilt—again and again, as will be seen—under the supervision of the original builder and bridge company president, Chauncey Thomas, who went to work, got the job done, and opened the bridge for operation at a cost of $4,000. The bridge treasurer, who hadn't yet paid any dividends to stockholders, couldn't pay this bill; he promised to pay the unhappy Thomas at a later date. At about this time Thomas was voted out of the presidency and James E. Gardner took over in his place. Gardner soon passed away and his place was taken by another stockholder, Napoleon B. Johnson. The bridge functioned adequately under Napoleon Johnson, and he remained the president of the Barryville and Shohola Bridge Company for a number of years. Then, on January 1 of 1865, a disaster occurred when the bridge's deck, overloaded with several heavy mule-drawn wagons, caused the cable to snap and the bridge to collapse into the river. The wagon operators all survived the icy river water but three mules were not so fortunate.

The bridge company was now in bad shape, financially, and getting worse. In addition to the bridge's destruction, President Napoleon Johnson had borrowed money for the bridge company that could not be paid back. And the company now had no funds with which to repair the bridge. After some hesitation, a sheriff's sale was held for the bankrupt and partially destroyed bridge. Former president Chauncey Thomas was the high bidder at the sale and, for only $1,979, became the sole owner of his old bridge. The Barryville and Shohola Bridge Company ceased to exist. More money was required for the extensive repairs, but they were done and, in addition, another pier was finally installed to increase the bridge's stability. The ferries, in the meantime, had gone back in

57 — The Barryville-Shohola suspension bridge was built in 1866 and lasted until 1941. Photo courtesy of John and Dorothy Bade.

operation to temporarily fill the gap while all this work was being done.

Finally, in the fall of 1866, the rebuilt bridge, still a wire-rope extension span, was completed and reopened for business. The cost this time was, again, $4,000. Chauncey Thomas remained owner of the span for the rest of his life, and this bridge, under him, became a prosperous money-maker and brought exceptional growth to the area. An 1872–73 booklet published by Sullivan County, in New York State, tells us that at this time, in addition to the fine bridge, "Barryville contains two hotels, two churches, four stores, three blacksmith shops, one wagon shop, a dry-dock for repairing canal boats, a grist mill, a stone quarry, a fine public school, and forty-seven dwellings. And the D&H Canal which runs through town has its own blacksmith shop and a superintendent's office, and, of course, docks for loading and unloading."

The village at the other end of the bridge, Shohola, Thomas's home town, had a fine railroad freight station, Thomas's store, several mills, and many substantial homes, but the citizens living on this side of the river did their wining, dining, and churchgoing in Barryville. Prosperity had indeed come to the area. Chauncey Thomas passed away at his Shohola home on October 5, 1882.

Thomas died without a will, and his extensive property was equally divided among his many children and grandchildren. A family friend and Shohola businessman, Stephen St. John Gardiner, was named the administrator of the estate. In this capacity, he purchased bridge corporation stock from Thomas's willing family members, who didn't want to be bridge owners, and Gardiner soon became the new and majority owner of the bridge stock.

This third bridge, the one put up by Chauncey Thomas in 1866, was apparently a

58 — A closer look at the one-lane Barryville-Shohola Bridge. Photo courtesy of John and Dorothy Bade.

considerable improvement over the previous structures. And Gardiner, when he took over, did some major renovation to it; he replaced the old cables with new ones of higher capacity, added new stringers, and laid a new deck floor. He even installed a new and more secure bridge railing. This substantial and cared-for bridge lasted well into the present period. It survived nicely the flood of 1903 and the upriver ice flood in the early spring of 1904.

The twentieth century saw many changes in the fortunes of the area. By the end of the nineteenth century the canal was out of business, as were the lumber and mining companies. Local prosperity, however, continued to grow. The area attracted many tourists as a summer resort, with several hotels and boarding houses built for the visitors. And at the same time some city folk built summer homes in the area, thus becoming local taxpayers without sending their children to the local schools in the wintertime. The now antique bridge that crossed the beautiful Delaware River here was a local attraction to these out-of-town visitors.

On January 20, 1923, the bridge was purchased by the Joint Bridge Commission of Pennsylvania and New York at a cost of $22,789.11, paid to the private owners. The bridge was then converted into a toll-free span.

The narrow, single-lane, old structure, now owned by the Joint Bridge Commission, served the citizens of Barryville and Shohola well into the twentieth century as a toll-free tourist attraction and asset. But old age was winning out. In 1939, the poor condition of the antique bridge caused the authorities to close the span to traffic until something could be done. After almost a year, some repairs were made to enable light car traffic to use it. Then, in early 1941, the Whittaker & Diehl Company was hired by the Joint Bridge Commission to construct a new and modern two-lane steel-and-concrete bridge for an impressive $174,300. On the Pennsylvania side of this new structure, a tunnel was built to carry bridge traffic under the Erie line to avoid stoppages and accidents here. The

59 — The present-day (2002) Barryville-Shohola Bridge, built in 1941. Photo courtesy of Samuel Sherrer.

new bridge was located slightly down-river from the old structure, and this enabled the old span to remain in service until the day the new bridge was opened for business. The new bridge went into operation just days before our country entered World War II. Such peace-time construction and reconstruction activities were soon stopped during the war; local citizens got their new bridge just in time.

The bridge structure at Barryville-Shohola survived World War II and several severe river floods totally undamaged. The new bridge, the wide and handsome two-lane structure, stands today on the banks of the beautiful Delaware. And it looks as new and impressive as it did on its first day of operation.

Life today in Barryville can still be exciting. The Indian League of the Americas hosts several activities here in the summer. Barryville also has a great canoe regatta, with races down the river from Barryville to Port Jervis, which draws crowds all along this part of the river. And on the Pennsylvania side of the river in the busy village of Shohola, the town's volunteer fire department holds an exciting carnival on the third weekend in July, which never conflicts with the Barryville schedule. Summers are fun here. Unlike some of the old villages along this upper river, Barryville and Shohola are still active communities.

TWENTY-TWO

The Lordville-Equinunk Bridge, 1870

In the old days, before a bridge had replaced the ferry at Lordville on the upper Delaware, the pedestrian who wanted to cross was offered, by the ferryman, a novel and exciting trip. The passenger was placed in a large basket hanging from a cable and then pulled across the river, quickly and safely. If he had a horse, it was put in a scow and poled across.

If this story is actually the truth, this service was probably the work of the area's prominent ferry owner, George Lord. He had received, in 1850, a license to operate a ferry on the Delaware River between two villages now named Equinunk, Pennsylvania, and Lordville, New York. Soon, the Erie Railroad made a regular stop across the river from Equinunk to utilize this ferry service. When this nameless New York settlement got a post office in 1854, it was christened with a biblical name, "Elam." A year later, Alva Lord was named postmaster of Elam, and the name of the village was changed to another biblical-sounding name, Lordville, in honor of his father, John Lord, the village's most prominent citizen. There were several other notable Lords living in Lordville as well.

This scow-ferry river crossing, like most other ferries up and down the Delaware, was eventually outgrown. Soon, local residents, including John Lord and his son, Alva, encouraged the construction of a bridge. John could only give moral support; he was busy taking timber rafts downriver to Philadelphia. In his seventy-ninth year, he made two raft trips downriver and then passed away. He willed his interest in the bridge to his favorite grandson, John Henry Lord. It was in consideration of grandfather John and the other raftsmen on the river that the piers of some of the Delaware bridges were designed to be farther apart than usual, so that the large, difficult-to-steer timber rafts could pass under them at high water and be able to fit easily and safely between these piers. Some of the later suspension bridges, especially on the upper Delaware where the river wasn't as wide, had no piers at all; Lordville was one of these.

The actual leader of the bridge-building organization and the principal stockholder in the newly formed Lordville-Equinunk Bridge Company was Alva Lord, John's son. And during most of the years of his life that he shared with the bridge company, he was its president and treasurer.

Alva Lord and other bridge supporters in the area were staunch admirers of the Delaware's outstanding bridge builder, John A. Roebling. This founding father, John, had recently died, but his bridge-building organization was now in the hands of his competent son, Washington Roebling. The Roebling firm was a thriving and busy concern at the time the Lordville citizens were planning a bridge; Roebling was working on the Niagara River, erecting a major bridge that would cross this river and connect Canada with the United States. Nevertheless, time was found to provide the engineering for the Lordville-Equinunk span under the supervision of Roebling's chief engineer, E. F. Harrington.

Harrington liked his new job and he liked the bridge owners, as well. He referred to the local bridge company leader, Alva Lord, as "The Lord." The new span, then, was built under the supervision of a Roebling superintendent, and much of the technical work was

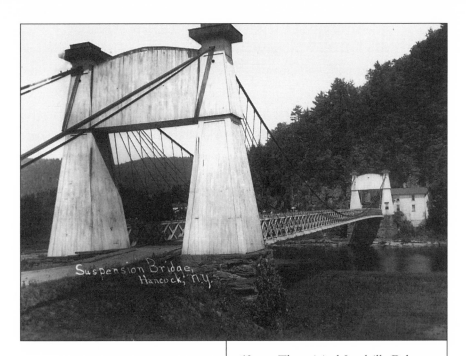

Suspension Bridge, Hancock, N.Y.

done by the Roebling Company. The structure was, after all, a wire suspension bridge with wooden towers, a Roebling specialty. On New Year's Day of 1870, Alva Lord had a special celebration—his bridge was completed, its wooden towers were painted, and his toll collector was on duty. It was opened for business. Happy New Year!

60 — The original Lordville-Delaware Bridge, a single-lane suspension bridge with wooden towers built in 1870 and destroyed in the flood of 1903. Photo courtesy of the Archives of the Delaware County Historical Association.

The bridge's early life was a busy one. It took over the ferry's business, but it soon did much more than that. By 1889, the shipments of milk from the Pennsylvania side of the river were crossing the bridge for delivery to New York markets. By 1893, a modern creamery had been established at Equinunk and it produced a large volume of bottled milk, which was hauled across the river to Lordville, then promptly shipped on the special Erie Railroad "Milk Train" downriver to the big cities. At the turn of the century things at Lordville, and across the river as well, kept improving. In May of 1902, the Hancock newspaper reported, "Lordville with two hundred inhabitants has a bridge that connects it to Equinunk, from which place large quantities of blue stone are shipped, as well as acetate from three large acid factories. And, of course, farm produce, all of which are shipped from here to various markets." Some of these markets were as far away as New York City.

This hard-working bridge needed regular maintenance and repairs, and apparently they were provided. In August of 1879, that newspaper from nearby Hancock reported, "The Lordville Bridge is receiving a fresh coat of paint and a large addition to the toll house." And in May of 1882, "Extensive repairs are being made upon the Lordville Bridge. Travel across it by teams of horses, after being suspended for three weeks, has resumed." In April 1895, the same newspaper reports a river flood. "At Lordville, one hundred feet of the Erie switch was washed away." The same issue describes other local damage, including that inflicted on the bridge. Mr. Harrington returned and made recommendations for repair to Alva Lord. Mr. Lord soon had a large crew working on the battered but still operating toll bridge. Shortly after this, the repaired span received another coat of paint, this time of a more modern color. "The bridge is being painted red, which adds greatly to its appearance," Mr. Harrington noted. Things couldn't be better. But the October flood of 1903 changed all this.

This catastrophe-to-come, due on the 10th of October, 1903, was preceded by heavy rainfall on the 8th and 9th of that month, causing the river to rise considerably. Late in the day on the 9th, the Erie Railroad's "Milk Train" entered the village of Lordville and was warned by the authorities to proceed no farther; trackage just downriver had been swept away. Train conductor Harry Lamb stopped his train's forward movement and parked on the tracks to wait. Soon the water level rose and then covered the tracks where he was parked. He then decided to back up to the high ground overlooking the village. Harry Lamb and his train spent the night parked on the hill, joined there by a passenger train. The next morning, the train crews and passengers witnessed the complete loss of the Lordville Bridge as it was swept away down the turbulent river. Part of the bridge, with its floating towers, reached Callicoon at about noontime. On October 14, the local Wayne newspaper, the *Independent*, concluded its story of the disaster with, "The loss is a very great public calamity. The personal loss falls heavily on Alva Lord, as he is reported to be the principal stockholder."

The Lord family, led by Alva, hired a prominent bridge-building concern, the Oswego Bridge Company, which had built several bridges by this time, including an iron bridge at nearby Narrowsburg that lasted until 1952. This firm soon got underway with the construction, erecting on the old piers an eye-bar chain suspension bridge with a 347-foot span having two full lanes for vehicles. The eye-bar chain Oswego used to support the bridge was a substitute for the wire cable that Roebling always used. Perhaps this was why Roebling didn't get a second chance; Lord wanted to try something different. The new structure was opened for business on June 4, 1904.

In the middle of the new bridge's construction, the owner and founding father, Alva Lord, passed away. Just about the time of the flood he had had a stroke, resulting in a creeping paralysis that ended his life on March 29, 1904.

The new bridge and its new owner, Alva Lord's young widow, Ida, and the old Erie Railroad all served the area well. Ida soon remarried. Her new husband was a widower named Walter Lambert, and he, too, became active at the bridge. President Ida served the bridge as both its chief executive and toll collector—with her new husband's assistance on the latter job—for many years.

The Erie line had the fine station in Lordville and served that community and surrounding villages, including Equinunk at the

61 — This suspension bridge replaced the original bridge in 1904 and lasted until 1986, when it was replaced with the present bridge. Photo courtesy of the Archives of the Delaware County Historical Association.

other end of the bridge. In the early years of the twentieth century the village of Lordville had, among its other attractions, seven stores and two acid factories . . . and its new, privately owned Delaware River bridge. It was a busy place. And the railroad was now bringing tourists—city folk—to the lovely upper Delaware Valley. Many of the wealthier visitors were soon building luxurious summer homes in the area. A few such homes still serve this purpose.

But privately owned Delaware River bridges were rapidly becoming a thing of the past. Government agencies were established in the concerned states to purchase these bridges, at a fair price, from their private owners and to reestablish them as government-owned and government-regulated—and toll-free—structures. Changing something from toll-charging to toll-free was a popular activity. As early as 1904, the Pennsylvania legislature was considering such a move for its Delaware River spans but was premature by about twenty years. In 1930, the Lordville-Equinunk Bridge was purchased by the New York–Pennsylvania Joint Interstate Bridge Commission for an impressive $26,000, payable to its principal stockholder, Mrs. Walter S. Lambert, Alva Lord's now somewhat older widow. Of all the Pennsylvania–New York Delaware River bridges, only two, by this time, remained privately owned—the Kellam Bridge at Little Equinunk and the Lackawaxen Bridge, an early product of John A. Roebling.

Under the new, government ownership, the operation of the Lordville-Equinunk Bridge looked to be uneventful. Maintenance was part of a regular program for the 1904 span and that was done—regularly. The notorious flood of 1955 was centered downriver and the river valley upstream above Callicoon was not even considered in the flood area. The New York City reservoirs had been created at midcentury on the east and west branches of the Delaware and the river's water sent overland to New York City rather than down the Delaware. This greatly decreased upriver flood threats, but the more common problem, old age, was unavoidable.

In February of 1984, the eighty-year-old span began to show symptoms. On that day, soil under one of the downriver abutments was washed downstream; the bridge then tilted and one of its towers twisted and then collapsed to the deck; the deck itself was soon sagging, drastically. The bridge was closed at once.

The New York–Pennsylvania Joint Bridge Commission was faced with three options: to repair the old bridge, to tear down the old bridge and build a new bridge, or to tear down the old bridge and not replace it with anything. The two states were long in making up their minds. Population in the area had thinned out and the need for river crossing here was not as great as it had been a century earlier. Unfortunately, the closest alternative was the Kellam Bridge, a small, one-lane structure that couldn't

62 — The new Lordville Bridge opened in 1992. Population is not as dense as it used to be, and bridge traffic is light. Photo courtesy of the author.

handle the additional traffic. The nearest two-lane bridge was the Callicoon span, which was many miles downriver, too far to be a substitute for users of the old Lordville-Equinunk Bridge.

The first decision made by the Commission, the fate of the old bridge, took over a year and a half to be acted upon; on November 24, 1986, at 10:20 a.m., the demolition got underway, and by day's end the bridge was gone. For the next four years, the cost to replace the bridge increased and a recession added to the concern. But local residents wrote letters to their politicians asking for a new bridge; some even contributed money to help pay for it . . . and they finally got their way.

The new bridge's general contractor was a firm named Barry, Bette, and Led Duke, but known to everybody, for obvious reasons, as BB&L. The parent company was a big-time Fortune 500 business, but its bridge division unit was small, down-to-earth, and competent. The bridge builders arrived on the job in May of 1991. They worked efficiently and quickly, but with some time off for an inclement winter, and the job was completed officially a year later, in 1992.

Celebrations were held and crowds of local people attracted, but since then, the thinly populated countryside has gotten back to normal. Judging from the amount of heavy traffic the new span deals with today, it should live a long and uneventful life.

TWENTY-THREE

The Pond Eddy Bridge, 1870

Early in the nineteenth century the Delaware and Hudson Canal was constructed and ran along the New York side of the Delaware River from Lackawaxen, hugging the river's eastern shore until it reached Port Jervis, where it then left the river's valley, turning eastward toward its ultimate goal, the Hudson River. A little later in the century, the blossoming Erie Railroad bought land and laid track, starting upriver on the New York or eastern side of the river, but then crossing over into Pennsylvania at Lackawaxen. So downriver from Lackawaxen, the Erie Railroad ran on the Pennsylvania side of the river, the D&H on the other side; together they brought a lot of business to the area.

In the early days of the canal, 1835 to be exact, a local family established the first hotel at Pond Eddy, located comfortably between the canal and the river. At first, the place was named the Sportsmen's House and catered to the new businessmen in town. This village soon became a busy stop for canallers, timber rafts-men, and their suppliers. Before long, the progressive town fathers of Lumberland, the New York township in which Pond Eddy was located, decided that a bridge should be erected here, thereby joining their section of the canal with the Erie Railroad station across

the river. This would make the rail ship-
ment of products, at the beginning
mostly lumber, slate, bluestone, and a
variety of farm goods, an easy process.
So, in the riverfront settlement called Pond Eddy in Lumberland
Town, in 1870, a bridge was constructed with local taxpayers'
money, and the owner was the Town of Lumberland, a unique
arrangement.

The structure was a handsome wire-rope suspension bridge, a
Roebling model, and built, some say, with the personal help of Mr.
Roebling. The construction was carried on under the immediate
supervision of an obviously competent local man, James D. Decker.
Decker was also the Sullivan County sheriff at the time and had
twice been elected town supervisor, or mayor. He lived within
easy walking distance of the bridge site, so close that for years
some local people referred to it as "Decker's Bridge." The bridge
was 521 feet long and 12 feet wide, spacious enough to handle the

anticipated traffic. The structure stood 31 feet above normal water level, which kept it safe from floodwaters for most of its life. The township also built a smaller, arched bridge so that people and wagons could cross over the canal and its towpath next to the river to get to the Delaware span.

Some historians believe that this larger bridge, from its beginning, was toll-free, at least for Lumberland's tax-paying residents. What actually seems to have been the case was that the township leased the span, and its tollbooth home, to private individuals who charged tolls for all users, including the local citizens whose tax money had paid for the bridge in the first place. Municipal bridge records tell us that John Main was, in the early days, the bridge's tenant and toll collector—he rented the span from the town and pocketed all the tolls he collected. But during long periods in the bridge's life, the town could not always find a willing bridge tenant, and township employees operated the span instead. Under these conditions, no tolls were charged and the crossing was free. Eventually, the tollhouse, no longer needed, was moved away from the bridge and converted into a private home. It still stands in town.

The new bridge crossed the river from the New York town to a little settlement in Pennsylvania, at the time named Flagstone, which, after the span was built, became Pond Eddy, Pennsylvania. Almost immediately, the two Pond Eddies began to grow and prosper. Pond Eddy, Pennsylvania, soon got its own railroad station, and then a fine general store. The riverfront bridge location on the New York side prospered even more so, and by 1880 Pond Eddy, New York, had a Methodist church, two stores, a hotel with restaurant, a telegraph office, and eighteen homes. The hotel with restaurant was the same structure built well before the bridge and called, then, the Sportsmen's House. It now had new owners and was named the Riverside Hotel, a prime stop here for travelers. This brought additional traffic to the bridge.

The Pond Eddy Bridge seems to have served the community well and made money for the citizens of Lumberland. In its early years there was growing commercial activity in the area, and weather in that period of the bridge's life was generally friendly. This good fortune didn't last.

During the last quarter of the century the Delaware and Hudson Canal began to decline, mostly because of railroad competition, until finally, in 1898, it went out of business. And this loss hurt the business activity in the area. The Erie Railroad Station on the Pennsylvania end of the bridge had no roads going elsewhere in the state and its business declined also. One historian, John W. Johnston, who lived in Pond Eddy, New York, in its heyday, returned there in 1900 and had this to say in his book *Reminiscences:* "Now the evident marks of dilapidation and decay appear in almost every part, and tend to cast gloom around the heart of one, like myself, so familiar with its residents and its industries in the brighter years."

One other significant thing happened in 1900 relating to the Pond Eddy Bridge. Its fine local builder, James D. Decker, passed away in July of that year at seventy-seven years of age. He had lived long enough to see his neighborhood and his bridge decline somewhat. To compound these difficulties, the handsome Pond Eddy Bridge was hit hard in the October flood of 1903, the worst in the area's recorded history, and was totally destroyed. Damage was also done to the railroad and to homes and businesses along the river in this area; only that fine old Riverside Hotel survived in good shape. The bridge-owning town of Lumberland faced, reluctantly, the high cost of new bridge construction and the sharp decline in bridge business.

Nevertheless, the original bridge was finally replaced by Lumberland Township. The bridge builder hired for the job was the Oswego Bridge Company. This firm produced a two-span, all-steel, but still only one-lane, structure. The cost for the job was

64 — This steel bridge replaced the first bridge in 1903. By this time the area had become a tourist center. The bridge is still in use. Photo courtesy of Frank Schwarz.

$28,900. And so, Lumberland Town continued as the bridge's owner, though reluctantly. The lumber produced on the New York side of the river and the stone mined on the Pennsylvania side were both used up, and this activity soon disappeared altogether.

But from the early years of the twentieth century, tourists visiting the area increased in number and made up for the loss in the areas of agricultural and small industrial businesses. In the early 1900s, the railroad brought increasingly large numbers of summer guests. Soon there were tourist homes and hotels along the river to accommodate them. Some of these seasonal visitors purchased summer homes, and several city families became year-round residents. And although that business has now declined, evidence of it still exists.

When the Joint Commission of Pennsylvania and New York came into being in the 1920s, Lumberland Township eagerly offered its bridge for sale to this bi-state governmental organization, but the Commission turned it down. The bridge was already free and receiving adequate maintenance without the Commission's spending money to buy it. And almost no Pennsylvanians lived near that end of the bridge to benefit from the outlay of Pennsylvania money. But the township was anxious to get rid of the maintenance responsibility and expense, and wouldn't take no for

65 — The Erie Railroad station across the river brought visitors and workers who walked to town over the bridge. Photo courtesy of Frank Schwarz.

66 — A mansion of yesterday near Pond Eddy on the Delaware. Photo courtesy of the author.

an answer. Finally, Lumberland officials led by Mayor Edward Bisland, who was a friend of Governor Pinchot, made the Commission an offer it could not refuse; they offered their bridge to the Commission for a firm price . . . of $1.00. At about the same time, the Narrowsburg Bridge just upriver was also purchased by the Commission for a respectable $55,000. Everybody on both bridges was happy and satisfied . . . so it was said. Both sales took place in 1926. The Pennsylvania–New York Joint Commission became, and still is, the owner of both structures.

The bridge has survived well since then. The Commission has maintained the bridge, used today by only a limited number of crossers, the few who live at its western end and some outdoorsmen. Much of the acreage immediately across the river in Pennsylvania, at the bridge site and inland, is now a secluded state fish and game homeland. Though the flood of 1955 brought threatening high water and considerable damage along the river's banks, the steel bridge survived, battered but intact. And, in more recent high water episodes—in 1967–68, for instance—the bridge survived unscathed.

But the bridge is getting older and there is some talk by the Joint Bridge Commission about a new structure, just slightly upriver. The present old, one-lane bridge cannot handle modern, heavy-weight fire engines, a condition that concerns the residents living on the Pennsylvania side, and the firemen, too. A new bridge here today would cost a little more—quite a little—than the previous spans built by Lumberville Township, and there is some local objection from the area's hard-working taxpayers. But for many others, it is hard to turn one's back on this pleasant past. So, time will tell.

TWENTY-FOUR

The Little Equinunk Bridge, 1889

Before there was a bridge over the river between Stalker and the Hankins area, there was a ferry on the river at this place. The ferry's privilege was granted to William T. Kellam by act of the Pennsylvania legislature on March 28, 1860. It was called Kellam's Ferry and it operated here for twenty-eight years, until the bridge replaced it.

The private organization that was formed to have the bridge built and to operate it afterward was the Little Equinunk Bridge Corporation. At its first meeting, on April 30, 1888, there were ten stockholders in the corporation. One, Joel G. Hill, a Civil War veteran, local landowner, and soon-to-be successful and popular state politician, became the first president of the bridge company; Franklin Holbert was the first vice-president; and David L. Kellam, William's brother, became the company's first secretary. Two other members of this original group were also Kellam boys, J. R. and H. P. Kellam.

The very first Kellams to settle in this village were their father and mother, Jacob and Hannah, who arrived in 1818. The father became a successful miller and landholder in the area. William was his first son, David was his second, and J. R. and H. P. Kellam were their younger brothers. David and his two younger brothers, had, like Joel Hill, served in the Civil War. David apparently also had some construction experience, for when this group was looking for a builder to construct the bridge, he bid for the job and got it. David Kellam was so confident that, before his low-bidder status was confirmed, officially, he started to work on the bridge. His bid price was $9,000 and he was, indeed, the low bidder. He actually went over this amount by $341.37. This sum was an extra charge for the construction of a railroad dock on the New York side of the bridge to accommodate anticipated—and soon realized—loading and unloading of freight trains of the New York, Lake Erie & Western Railroad, by this time known simply as "The Erie Railroad." The main railroad station in this area was at nearby Hankins, but the Erie quickly established a flag stop and siding at the new bridge. This bridge and the railroad siding are still there.

The construction of the Little Equinunk Bridge, a one-span, wooden, wire-rope suspension structure, painted red, was started in 1888 and finished the next year. Although the bridge was completed and in use by 1889, the vital tollhouse was not built until September of 1891. Somehow the able toll collector did his job without the shelter. The toll building was more than just a toll-booth; it was the home of the collector and his family. The first toll collector, unnamed, in addition to receiving living quarters, was also paid $20 a month, and this rate seems to have prevailed. There were a few husband-and-wife teams who worked together as toll collectors, or sometimes took turns. Mr. and Mrs. Pelham were one hard-working pair; and Mr. and Mrs. J. D. Bailey established a toll-collecting record—they worked at the job for nineteen years until Mrs. Bailey passed away, but her widower carried on at the job without her. Not as popular as the Baileys was toll collector Jacob Barrager, who did the job confined to his wheelchair. He kept a loaded gun in the tollhouse to enable him to stop people who might attempt to cross without paying.

67 — Little Equinunk Bridge, sometimes called Kellam's Bridge. This is the original structure, built in 1889. Photo courtesy of the author.

A notice of the rates to be charged was posted at the entrances to the bridge, and tickets were sold here. A single ticket for a one-way trip with a team of horses and vehicle was 25 cents; for a round trip, only 35 cents. A mailman on the job had to pay, but only 15 cents for a round trip, and that included his horse and vehicle. Using the railroad dock for loading stone, lumber, grains, and so forth was 50 cents for each rail car loaded. If one hundred tickets were purchased at one time by a customer, there was a discount.

Business was good and getting better. Dividends were paid to stockholders twice a year almost every year, although on a few occasions of high repair expense (1896 was such a year), dividends were not paid.

David Kellam served his bridge well for a number of years, both as corporation secretary and bridge builder, so well that the span was named, unofficially, "Kellam's Bridge." He died in 1895 at the age of sixty-two. The funeral was held in his hometown of Little Equinunk, and he was buried in a cemetery in the shadow of his bridge.

As time passed, things changed somewhat. Toward the end of the century the name of Kellam became less common, and the Hill family began to take charge. When the bridge was erected, Joel G. Hill was the first president of the Little Equinunk Bridge Company. Years later, when the bridge was becoming the property of the bi-state agency, Joel Hill's son Harris was the company's last president, and his son John was its last treasurer. By this time Joel Hill had passed away, his death the result of burns suffered in a fire at his home.

Occasional high water did not seem to be a major problem here. During the great Pumpkin Flood of 1903 the nearby Lordville Bridge was swept

68 — The Erie Railroad ran close by on the New York side and, in the old days, brought business. Photo courtesy of the author.

away, a total loss, but the neighboring Little Equinunk span apparently avoided major damage. Bookkeeping records for

69 — The bridge is one-lane, but nobody seems to notice. Photo courtesy of the author.

this bridge show profitable years with dividends paid to investors, except for 1903 and 1904—this last was another stormy year for some bridges on the upper river. Other flood years along the upper Delaware were 1922, 1936, 1942, 1955, and 1972, but they did no major damage to the Kellam span. This bridge was considerably higher above normal water level than many of the others and was located in the center of river flats, which kept the water level low during periods of heavy rainfall. Just upriver at the Lordville span, the river's passage was narrow and twisting and the riverbanks steep from the river's edge and upward. Such a configuration caused occasional and quick flooding in the Lordville area. When neighboring bridges were closed for flood repairs or rebuilding, their customers used the old but dependable Little Equinunk Bridge to cross the river. Today, with the twin reservoirs nearby, one on each branch of the divided river above Hancock, some water is removed from the upper river and, apparently, so is the threat of upriver floods.

The acquisition of this bridge by the Joint Bridge Commission was no easy matter. The *Hancock Herald* of October 29, 1932, tells us: "The bridge was finally taken over by the Joint Bridge Commission. Ever since the Free Bridge Bill was passed by the New York legislature a decade ago, the structure has been a source of much wrangling and litigation."

The stormy deal was finally concluded in December of 1928 at a price of $15,000, not bad at the time for an old, single-lane span. The last officers of the Little Equinunk Bridge Company were all prominent men of the Hill family. No Kellams were involved in this final sale.

In 1936 the Joint Commission owners made some major repairs on the bridge. It now has a steel grillwork floor in place of the old wooden deck. But to this day the span is still a single-lane, single-strand structure. Since this change of ownership, bridge business in the area has decreased sharply, and apparently permanently, and this special structure is not overworked. Not only is the bridge still a single-lane span, but so, too, is most of the twisting road serving it. The bi-state owners, however, treat it well but gently, and hopefully it will last a while longer.

TWENTY-FIVE

The Callicoon Bridge, 1898

Long before there was a bridge over the Delaware at Callicoon, many of the local townsmen used their river to take huge timber rafts downstream. At first, they took the rafts all the way to Philadelphia to sell them, but later, when lumber mills opened for business at the nearer villages of Phillipsburg and Easton, that area became a last stop for Callicoon raftsmen. The champion Callicoon timberman of that time was Elias Mitchell, who proudly accepted the title of "Deacon." In the old days, all the timber raftsmen worked hard and dangerous twelve-hour days on their trip down-river, but the younger raftsmen, and some of the older ones as well, spent part of their evenings ashore in local pubs or hotel bars, or, if they were spending the night in a private residence, as many did, time was often spent with the owner and his family, especially the owner's young daughters, having a good time.

The outstanding exception to this activity was the Deacon, Elias Mitchell, who followed a rafting routine totally different from that of his fellow raftsmen. He managed, with his raft and crewmen—and a full moon, to go from Callicoon to Easton in just half the time of his fellow rafters. If he left Callicoon early in the morning, he would reach Easton the next morning, bright and early . . . by traveling all day and all night without any stops along the way for rest or relaxation. He got the same amount of money for the job but gave only half as much of his time. Of course, the new Callicoon Bridge, when it came into service, saved time for everybody.

At the first meeting of an inspired group of Callicoon bridge supporters, held in February of 1886 at the local Everard House, the need for a Delaware River bridge in their town was emphasized. The nearest bridge to their community at this time was the old Cochecton-Damascus span ten miles downriver, too far away to serve the growing community of Callicoon. This meeting was reported by the local newspaper, the *Callicoon Echo,* and the bridge, thus, won even more support. But the passage of time before the bridge could be built saw a decline in interest; the bridge had to be approved by the states involved, land then had to be acquired on both sides of the river, money had to be raised, and a bridge contractor hired, all before work could begin. Unfortunately, it would be another ten years before this bridge was built.

The group's members and friends persevered, however, and a private concern, the Callicoon Bridge Company, was formed. The first president of this firm was Charles T. Curtis, whose descendants are still active in Callicoon. Stock in this corporation was available to the public at $25 a share and was sold to eager buyers. And when it had collected some money, the Callicoon Bridge Company purchased the property for the bridge—the New York section, at least—from a local citizen for $150. A generous Pennsylvania lady who owned the matching land across the river presented it to Curtis . . . at no charge.

In 1898, the Callicoon Bridge Company hired the Horseheads Bridge Company to actually build the structure, a suspension span. The price was $23,200. The Horseheads Company was owned by three brothers, James, Will, and E. Perkins, of Horseheads, New

York. The Perkins brothers constructed this span efficiently and quickly, for they had another Delaware River bridge-construction job awaiting them, at a place downriver from Callicoon, called Dingman's Ferry.

Local man Jacob Knight was the first toll collector at the bridge, and he occupied the bridge-tender's new house next to the Callicoon Bridge. He was a harness-maker by trade who was looking to better himself. He and his wife and children moved into the two-story tollhouse and he took his harness tools with him and did that business on the side, when he wasn't taking tolls. He also bought stock in the bridge company and wasn't shy about his ownership of a bridge. Soon, in 1906, he became one of the directors of the bridge corporation.

The Callicoon Bridge was opened to the public on January 4, 1899, at which time a party was held and, on that day, at least, no tolls were charged. Thereafter, the bridge was opened and operated for business and profit. In its first year,

70 — The original Callicoon Bridge, opened in 1899. After the storm in 1904 the bridge was raised. Photo courtesy of the Equinunk Historical Society.

71 — The first Callicoon Bridge after it has been raised. Photo courtesy of Robert Longcore.

with Jacob Knight on the job, $1,254.83 was collected in tolls. This was a promising beginning for the Callicoon Bridge Company. And the Erie Railroad ran on the Callicoon side of the river as far down as Narrowsburg. The new bridge was able to give good service at its Callicoon Station.

The bridge did well . . . for a while. It survived the 1903 flood, which was very destructive downriver; it destroyed four bridges upriver, as well. And a severe storm struck the upper river the next year, in March of 1904. After this early-spring, ice-filled flood, which caused some serious bridge damage, the Horseheads Bridge Company was hired again, this time to raise the bridge piers and deck, and the bridge-tender's house, to better enable these structures to avoid future floodwater and winter ice. This job cost the bridge corporation $2,849.

The only mishap for Horseheads on this job, according to the local Callicoon newspaper of October 19, 1904, involved one of their employees who "had the misfortune to fall from the scaffolding to the rocks below, a distance of about 20 feet. He miraculously avoided serious injury and was back at work in a few days." So even this event had no serious consequences.

This improvement solved the bridge's problems, and although earlier prosperity seemed to have faded somewhat, the bridge operated adequately into the twentieth century. Another major Callicoon business, timber rafting, also continued, to some extent, into the twentieth century, but one trip down to Easton had a tragic outcome. The raft owner was Art Mitchell, related, no doubt, to "Deacon" Elias Mitchell, the famous Callicoon raftsman of an earlier generation. Art Mitchell owned the raft but hired J. Skinner of Milanville, an experienced raftsman, to run it down to Easton for him. As the raft was racing downriver near Belvidere in New Jersey, it hit a Myers ferryboat crossing the river attached to a cable. The ferry held two cars and several passengers. Four people on the ferry, three women and a young man, were killed. The ferryman was blamed for the accident, but the experience seemed to mark the end of timber rafting for Callicoon.

The twentieth century brought some good fortune to Callicoon, too. On February 8, 1923, the states of Pennsylvania and New York paid the private owner, the Callicoon Bridge Company, $35,000 for the bridge. Tolls were immediately dropped from the operation, and the bi-state bridge corporation became the new owner. The Callicoon Bridge Company was no more.

The location of Callicoon on the official river map also indicates the most northern river area hit by the great flood of 1955. Callicoon was the place. Native folk remember the high water and the river areas that were flooded. But the bridges in the northern New York State area were, for the most part, unharmed. The major damage was downriver in New Jersey, where four spans, three of them steel structures, were totally destroyed. Floods, then, did not cause the downfall of the Callicoon Bridge. Old age did.

As time passed, traffic began to increase again, and so did the weight of vehicles. Heavier trucks, buses, and cars began to take

72 — The new Callicoon Bridge, built in 1962, a handsome structure. Photo courtesy of Samuel Sherrer.

a toll on the old river span that was born in a different, lighter age. It was decided by the bi-state bridge organization that a new span had to be constructed; the old one was beyond repair. The new structure was started in 1961 and finished in 1962 at a location slightly downriver from the old span. The old bridge, thus, remained in operation during the building of the new structure, and the local toll-free users were content.

The new bridge was built by a most reputable corporation, the Binghamton Bridge and Foundation Company. This structure is much larger and sturdier than the old one. The new bridge was subjected to a flooded river in the late 1960s and to high water several times thereafter, with no negative results. A new addition, the creation of two reservoirs, one on each branch of the river above Hancock, attracted attention. They were to supply faraway New York City with water for those good and very numerous citizens. This made a marked reduction of the river's water level upriver, and since the creation of these two reservoirs, bridge flooding upriver from the Cochecton area, including Callicoon, Little Equinunk, and Lordville, seems to have decreased.

This fine and modern bridge structure, still toll-free, is the pride and joy of Callicoon residents and a favorite crossing for travelers in either direction. In the summer, the bridge gets some extra use; the village is the center of vacation activities in the area. In early June the townsfolk hold a three-day festival of storytelling, folk music, and traditional crafts, which draws thousands of visitors from up and down the valley. And a month later the great Canoe Regatta, which begins far upriver at Hancock, ends in Callicoon with appropriate festivities for the canoeists and their fans.

The new river bridge, like the whole village and its citizenry, is up-to-date and leading an active life.

TWENTY-SIX

The Lackawaxen Bridge, 1900

The present-day suspension bridge crossing the Delaware at Lackawaxen, Pennsylvania, had a unique beginning. Designed and completed in 1848, this span was constructed under the supervision of the period's outstanding bridge architect and builder, John August Roebling. It was erected in an area with a prior claim to disastrous fame as the site of the 1779 Battle of Minisink and was part of the Delaware and Hudson Canal, twenty years after the canal opened for business, to improve the crossing time of the canal boats loaded with Pennsylvania anthracite coal and on their way to the Hudson River and thence to New York City.

This aqueduct was one of four built by Roebling to handle the canal's heavy traffic crossing a river; another was constructed over the nearby Lackawaxen River just before it joined the Delaware, and the last two were erected in New York State, one across the Neversink River at Cuddlebackville, and the other over the Rondout Creek at High Falls, very near the Hudson River. The largest of these four aqueducts, the Delaware River crossing, is still in use today after 150 years, but now carries travelers in automobiles across the river. John Roebling would later design the better known Brooklyn Bridge in New York but would die before the

job was finished; his capable son, Washington Roebling, actually built the Brooklyn structure.

For canal boatmen, this Delaware River span was a welcomed addition to the canal. Prior to its existence, canal boats had to enter the Delaware River at Lackawaxen and pull themselves slowly across the river with a ferry rope. The animals and crewmen, who normally propelled the craft, crossed the river separately in another boat. This was a dangerous and sometimes fatal crossing, especially in the spring, when brakeless timber rafts raced downriver in the high water.

In the initial construction of the canal, in 1827, the D&H Canal people built a dam across the river here, the first and last such dam anywhere on the Delaware. They built it to be sure that their boats would have enough water to make the crossing even during the dry season, when the water was low. The speeding and brakeless timber rafts shot over the top of the dam. Accidents caused by the dam and by the growing congestion of canal boat and raft traffic resulted in many financial claims by the raftsmen against the canal company, and the company paid large sums of money to the rafters. But the timbermen, when they complained vigorously about the dam, apparently were ignored. Shortly thereafter, according to the April 18, 1829, *Easton Argus*, "A public meeting was held by those who considered themselves aggrieved, and a representation sent to the managers that if the obstruction was not immediately removed, they would remove it by force. The request was not complied with. The group of raftsmen proceeded to the dam, blew it up and tore away some eighty feet of the dam, thus clearing an adequate passage for their rafts."

The D&H Canal executives wisely decided to maintain this opening for the rafts. They then undertook the building of the four water bridges, or aqueducts, over rivers on the canal's route so that their crafts would no longer come in contact with the dangerous timber rafts.

The Delaware River Aqueduct didn't result in great happiness among the timber raftsmen, for in addition to the challenge of locating the gap in the dam while approaching at a high rate of speed and without brakes, the haggard timbermen now had to dodge—or try to, anyway—the piers that supported the new Delaware Aqueduct that passed overhead. But at least the canal boats' crews were happier and safer . . . and the timber raftsmen's claims decreased.

The span, 535 feet long, was a suspension bridge partially supported by cables. This meant that fewer piers were needed to support the bridge and this made passing under the bridge easier, somewhat, for the timber raftsmen. The trough on the aqueduct that would carry water and the canal boats above the river was 20 feet wide and 8 feet deep. A plank towpath on the structure was built for the mules to walk on while pulling the craft across the

river. Both the Delaware Aqueduct and the shorter Lackawaxen Aqueduct were completed by the end of 1848 and were ready for use when the 1849 season opened. The new structures, engineered by Roebling, functioned perfectly, and to Delaware River loyalists these bridges were as good as anything that would later appear in Brooklyn.

The canal's coal volume increased greatly with the building of these aqueducts, especially the big one that crossed the Delaware River. In the year 1851, half a million tons of Pennsylvania coal were transported on the canal, and by 1856, just five years later, this amount had doubled to over a million tons. The company paid a dividend to its stockholders of 18 percent in 1855 and 16 percent in 1856. Mr. Roebling's structure contributed much to this prosperity.

But the good old economic philosophy of the United States, free enterprise, made this money-making activity attractive to competition and that is what the railroad was. The Erie Railroad could transport the canal's products faster . . . and cheaper. And other railroads wanted to get in on the act. That wonderful mule-powered D&H of yesterday was, by the end of the century, obsolete. The Delaware and Hudson Canal, after losing money for several years, went out of business in 1898.

The next year, 1899, a third party, the Cornell Steamboat Company, who through the years had worked with the D&H, became its new owner. This firm immediately put the former D&H property on the market, offering it to railroads as a first-rate route from mining and manufacturing areas of eastern Pennsylvania to the New York City metropolitan area. In its initial construction, in 1820, all of the canal route had been cleared and most of it leveled by the D&H; the final preparation for railroad use would be simple enough—"Fill the ditch, lay the track, blow the whistle."

There were several railroads interested in purchasing the old canal property and establishing a rail line on it. The two most deeply involved, however, were a team composed of former officers of the D&H Canal Company, who had recently founded the Delaware Valley and Kingston Railroad in New York State, and another group, established by the Pennsylvania Coal Company to operate the railroad in Pennsylvania and christened the Erie and Wyoming Valley Railroad. Both of these firms, in 1899, bought the canal property in their states from the Cornell Steamboat Company. They had formed a partnership that would include all the former canal property in New York and Pennsylvania. The Erie and Wyoming Company was in no way related to the Erie Railroad, already well established in the Delaware Valley. It was the Erie Railroad whose competent and inexpensive service in the valley had put D&H out of business and created a monopoly for itself in this area; the Erie Railroad had no intention of giving up its monopoly to this new organization. After eight years of struggle, legal and financial, the Erie Railroad won the battle. All schemes to bring a competitive railroad to the area were abandoned.

A new commercial activity in the area, one that succeeded, was the conversion of the Roebling Delaware Canal Aqueduct to a pedestrian and wagon bridge across the Delaware River. On May 12, 1908, Charles Spruks, a Scranton lumber dealer, purchased the aqueduct from its defeated owners, as well as the roadways leading to it on both sides of the river. Spruks spent a lot of time in this area on business; he was purchasing his timber and lumber in New York State and had to get it across the river to ship it on the Erie Railroad. He bought the bridge at a bargain price and then converted it from an aqueduct to a horse-and-wagon span in order to carry lumber, by the wagonload, across the river, and then ship it by rail on the Erie. And as an afterthought, new owner Spruks adapted the bridge to accommodate other river crossers. He constructed a gate and a tollhouse at the New York entrance and erected a railing as well. A few locals had been using the aqueduct, toll-free, when nobody was supervising things, but now, under Charles Spruks, toll-free crossing ended. It became a profitable sideline and then a profitable mainline river crossing.

In a 1914 edition of *Harper's Magazine,* writer Edward Hungerford had this to say about the new crossing: "It is a sturdy wooden structure, wondrously fashioned. For sixty-five years it has defied the fearful springtime floods down the Delaware—floods that have played havoc with more modern bridges. Today, it is itself a highway bridge of importance. And where the slow-moving coal barges once made their weary way, the autos and trucks now have a quick and easy flight."

Spruks frequently had heavy loads of timber hauled across on his bridge, sometimes pulled by a pair of horses instead of just one. The bridge handled the weight without a problem.

Spruks often bought additional land on the New York side of the river because of the high quality of the timber there. He spent a good amount of his time on the road for his lumber business and, for a while, had a local overseer named Andy Faye run the bridge business for him. One of the earliest toll collectors was Alonzo Smith, who lived in the four-room tollhouse with his wife and daughter. In the 1920s, another family, the Campbells, collected the tolls for Spruks.

Mrs. Campbell, Ella, soon had a nickname; she smoked a pipe and used a brand of tobacco called "Yellow Daisy," and her nickname among bridge customers became . . . "Yellow Daisy." Under the Campbells, the bridge hours were from 7 a.m. to 9 p.m. and the toll window was in the kitchen on the first floor of their tollhouse, a convenience at mealtime. Tolls for walkers in the 1920s were 2 cents; cows or pigs were 3 cents. The Campbells and many local residents, in the mid-1920s, observed a small private airplane fly under the aqueduct, just barely fitting between the stone bridge piers and under the bridge's deck. The pilot and his plane made it, and, fortunately, never tried again. The pilot was also christened with a nickname—"The Airplane Cowboy."

Zane Grey was the local Western writer, who, in the early 1900s, lived in a large and lovely home in Lackawaxen just upriver from the bridge. He frequently used the bridge to cross over and inspect property he owned on the New York side of the Delaware. Readers are welcome to visit his fine home today; it is now owned and nicely maintained by the National Park Service.

In November of 1930, after almost a quarter of a century of profitable bridge ownership, the now older Charles Spruks and his wife, Nettie, reached retirement age. They sold this bridge to the Lackawaxen Bridge Company, a branch of a big-city, privately owned, multi-bridge-owning firm, the Federal Bridge Company. This new owner seemed short of cash, and records of this sale indicate that the buyers gave the sellers $3 in cash and signed a $35,000 mortgage, held by the Spruks, to pay for the purchase. The Lackawaxen Bridge Company then began using its cash funds to modernize and increase the capacity of the newly purchased one-lane structure. This was done to enable the bridge to handle the heavier cars and trucks coming into use.

John H. Redding was one of the first toll collectors to work for the new owners; the tollhouse became his home. One of the other toll collectors in the 1930s was a capable woman, Mrs. Heflin; collecting tolls, by now, was an acceptable female occupation.

The positive growth was interrupted by a serious fire on the bridge on May 30, 1933. Anything that was wood—most of the superstructure—was destroyed on the Pennsylvania span and on the bridge next to it. The rest of the structure was saved. The bridge was closed briefly until the damaged sections were replaced. The cause of the fire was never investigated by any law enforcement agency, but it was the general opinion of the town's citizens that the fire was started by locals who wanted it totally destroyed and then replaced with a modern and toll-free structure.

Carl Draxler, a Lackawaxen toll collector, recalled that by this time most of the Delaware River spans had become bi-state-owned, toll-free bridges. He remembered that many of the unhappy local bridge users at Lackawaxen who resented paying a toll, instead of resorting to arson, took a slightly longer trip and crossed the Delaware downriver at the nearby two-lane, all-steel Barryville-Shohola Bridge. It was toll-free.

One of the longest lasting toll-collecting families was that of Jack Brower with his wife and three able-bodied children, all of whom did whatever work was called for, including collecting tolls. Members of this family made minor repairs, shoveled snow, cleaned and painted the bridge, and were polite to all the customers. The old tollhouse was lacking any indoor toilet or running water, but the Browers were supplied with a rope and pail to get their drinking and washing water from the river. This family worked hard for a number of years until 1940, when one of the children died unexpectedly, followed soon by the father, Jack. The mother and her one son (her oldest daughter had married and departed earlier) ran the bridge for another year and then moved away in the summer of 1941.

On June 1, 1942, Edward H. Huber, another lumberman from Scranton and the nephew of Charles Spruks, purchased the Lackawaxen Bridge Company and its bridge from the privately owned Federal Bridge Company. It didn't cost him any money, for he had inherited his uncle's $33,000 note for money owed by the Federal Bridge Company since it purchased the bridge.

Edward Huber did not plan on spending the rest of his life at Lackawaxen collecting tolls. Rather, he felt he could make a deal with the two states for them to assume ownership of this next-to-last privately owned bridge on the Delaware. But Pennsylvania, New York, and Huber couldn't agree on a price; owner Huber stayed a little longer than he had planned. He spent the next thirty-one years of his life operating the Lackawaxen Bridge.

New owner Huber didn't react negatively to his failure to make a deal with the two states. He established a high level of maintenance and safety on his bridge and also collected increased tolls to pay for its upkeep. Regular daily inspections were made by the toll collector, and a time during the winter's slower period was set aside for major repairs. Edward Huber soon became the hardest worker on the Lackawaxen Bridge . . . and in 1968, he got recognition; his old aqueduct was named a National Historic Landmark.

Huber kept up this pace until 1973, when, at age seventy-five, he put his bridge up for sale. At first he got very little reaction, at least from potential buyers. Historical organizations in the area and beyond, in some cases, advised him to be careful as to whom he sold it, to be sure it wasn't knocked down and replaced by a more modern structure. He listened to them. And he eventually found a buyer for his bridge.

The new owner of the bridge was Albert Kraft, a younger man with more modern ideas. He soon made a practice of hiring local teenagers as toll collectors. These youths lived at home. The first floor of the toll building remained as the office for collecting tolls, but the second floor was soon converted into a gift shop, an attraction for visiting tourists.

Kraft raised the tolls. The main vehicle using this crossing now was the auto; its toll was raised to 35 cents and then, again, to 50 cents. And for additional income, owner Kraft purchased the nearby Lackawaxen House Restaurant. Toll money was used to modernize the restaurant instead of maintaining the bridge. The vandalism that had become a problem for previous owners increased; the tollhouse was broken into several times. On June 2, 1977, a flat-bodied truck with a load of ties broke through the bridge deck and landed upside down in the abandoned D&H Canal, 30 feet below. There were no casualties, but a lot of labor and expense was required to replace the rotted wooden deck. Kraft closed the bridge and repaired the deck, but by the time this was completed, the now discouraged Kraft had lost four months of prime bridge time. With the period of slow winter business approaching, it remained closed.

At about this time, in the fall of 1978, the Upper Delaware Scenic and Recreational River organization was established as a part of the National Park System in the hope that it would be able to take a direct and active part in the salvation of the Lackawaxen Bridge. When the disheartened owner, Kraft, put the bridge up for sale, his most promising prospect was the National Park Service. This organization made an offer that Kraft did not want to refuse . . . but had to. He found himself in legal trouble over the bridge and truck accident of June 1977. This problem had to be straightened out for him to get a clear title for the sale of his property. It took about two years, but in May 27, 1980, the National Park Service became the new owner, with good title, of the Lackawaxen Bridge Company. The National Park Service paid the former owner, Albert Kraft, a handsome $75,000.

75 — The former aqueduct, now handling horses, wagons, cars, and trucks. As a vehicular bridge it is now over one hundred years old . . . and still open for traffic. Photo courtesy of Robert Longcore.

There was a lot of work to be done by the National Park Service, but it got underway almost immediately. John A. Roebling's fine structure was opened again in October of 1980, this time toll-free, but only for lightweight pedestrians and bicycle riders. A little more would have to be done before motor vehicles could cross again. In the meantime, in addition to bridge repair, work was done on nearby roads, outbuildings, even the tollhouse. Government money to pay for all of this was not always prompt in coming, but finally, in 1987, Roebling's wonderful structure was officially reopened. His birthday was June 12, and opening exercises were held on the next Saturday, June 13. Three thousand people attended. The final ceremony of the day was the reopening of the faithfully restored Roebling bridge, which, at that time, was 139 years old, the oldest, by far, of any such spans in the United States.

The National Park Service has established a headquarters one hundred feet or so upriver from Roebling's masterpiece, in the restored and spacious home of Zane Grey. These buildings and that special Roebling bridge have all been restored, accurately and functionally. John A. Roebling has achieved a rare American immortality; not bad for that foreign-born engineer.

TWENTY-SEVEN

The Milanville Bridge at Skinner's Falls, 1902

Long before there was a bridge crossing the Delaware at Skinner's Falls, the area's prominent family, the Skinners, produced capable rivermen. It was from Skinner's Falls that Daniel Skinner, in 1764, took the first timber raft down the Delaware River. He received for this the unofficial title of "Lord High Admiral," an honor that still shines. The Skinners' primary homeland seems to have been on the Pennsylvania side of the river at a place christened officially as Milanville, but known locally as Skinner's Falls. Just upriver from the falls, another Skinner, Milton L., operated a ferry between Pennsylvania and New York for a number of years. He and his brother, Volney, also operated a sawmill at the falls, on the Pennsylvania side. These two rivermen were prominent in the community.

It wasn't until the beginning of the twentieth century that it was felt appropriate by politicians that a bridge should be built at Milanville/Skinner's Falls. Soon, a private bridge company was formed to sell stock, build the structure, and then operate it for the general public. The firm was named the Milanville Bridge Company; its home office was in this little village called Milanville at Skinner's Falls on the Pennsylvania side of the Delaware. The official name of the structure, then, was the Milanville Bridge, but it was more commonly referred to, ever after, as the "Skinner's Falls Bridge." The company's first president was Milton L. Skinner, the same man who had operated a ferry at this spot on the river. The corporation received its state charter in late 1901, and hired the American Bridge Company to do the actual construction. In February of 1902, before much work could be done on the bridge's construction, an ice flood hit the area and immersed many riverside Milanville homes to the second floor. This and the strong objections to this span from nearby competitive bridges caused a slight delay.

Just three miles to the north of Milanville was located the Cochecton Bridge, and three miles south was the Narrowsburg Bridge. Both of these neighboring, privately owned bridges dated back to the early years of the nineteenth century, and the bridge owners objected to this close and, as they thought, unnecessary competition. John Anderson, the attorney for the Cochecton Bridge, made a special trip to the state capital at Albany to fight the building of this nearby structure. Possibly, it was the result of the valid objection by these two close-by bridge competitors that this modern steel bridge at Milanville was limited in its width so that it could handle only one lane of traffic at a time. People in a hurry would prefer to use either the Narrowsburg or Cochecton spans and avoid having to wait in line. This compromise decreased objections and soon the Milanville Bridge was under construction. The builder, the reputable American Bridge Company, finished the job in November of 1902 for a mere $14,000.

Initial tolls to cross this bridge, either with horse or auto, were a bargain when compared to such charges on either the Narrowsburg or Cochecton Bridge; 22 cents as compared to 25 cents. And a poor customer who could only travel on foot across the Milanville Bridge frequently was charged nothing at all, an unheard-of custom at Narrowsburg or Cochecton . . . or any other toll bridge. The

76 — Milanville, or Skinner's Falls, Bridge on the upper Delaware, built in 1902. Photo courtesy of Samuel Sherrer.

Milanville Bridge was also free to local ministers so that they could more readily serve their community. However, when a Baptist minister was obviously abusing the privilege, he was charged 5 cents for each crossing thereafter. The tollhouse was in Milanville on the Pennsylvania side of the river and for a while was operated by a husband-and-wife team, Mr. and Mrs. Albro Dexter.

Then, just two years after the bridge opened for business, that upriver curse, the flood of March 1904, did serious damage to the Milanville Bridge. Flooding of the river carried off, on a sheet of ice, the New York section of the bridge to nearby Skinner's Falls, where it ran aground. The Horseheads Bridge Company, with those hard-working Perkins brothers, was hired for the repair job; their price was lower than anyone else's. The Perkins men were able to do it cheaply by reusing girders from the damaged section to make the repairs. They charged the bridge owners only $7,000 for the job. The little bridge then promptly opened up again and led a somewhat busy but simple and trouble-free life.

The reopening of this bridge seemed to attract other business activities to the area. The Erie Railroad established a Skinner's Falls Railroad Station and also constructed a siding and a freight station for the Milanville Acid Factory, a local branch of the big Brant-Ross Chemical Company. The Skinner's Falls Creamery was established here, and, later, the Fullerbosne Dairy Company also moved in and built a creamery. Local farmers no longer had to take their milk to Cochecton. The Fullerbosne Dairy Company paid local dairy farmers 1 cent per quart for milk. A farmer crossing the bridge with his horse-drawn wagon paid a toll each way of 22 cents. These added businesses created new bridge traffic and a marked increase in toll income for the bridge and its owners.

One of the river's businesses, timber rafting, which got its start with Daniel Skinner in 1764, ended with J. Skinner, another Skinner's Falls citizen, in 1914. He had all the skill of his ancestor, apparently, but not the good luck. He was taking a timber raft downriver in high current when, as he neared Belvidere, the Myers ferryboat pulled in front of the timber raft and got hit. Four passengers aboard the ferry, three women and a young man, were killed, and the ferry operator was blamed. This tragedy, and the shortage of riverside forests by this time, ended commercial timber rafting on the Delaware.

To city folk, things looked good in the upper Delaware Valley. In the early and middle years of the twentieth century visitors from eastern New Jersey and New York were drawn, in growing numbers, to this beautiful and secluded river valley area surrounding Skinner's Falls. Milanville and nearby Cochecton had five luxurious tourist resorts in their vicinity that were busy, summer, spring, and fall. One such establishment erected a sign describing its rooms as "Large and Commodius." All this business was accommodated

77 — Milanville Bridge remains a single-lane span. Photo courtesy of the author.

78 — Bridge employees still keep this sign shining and legible, immortalizing these local giants of the past. Photo courtesy of Samuel Sherrer.

by the narrow Milanville Bridge, and its neighboring spans at Narrowsburg and Cochecton, and the Erie Railroad, as well.

The "free bridge fever" took over on this bridge as well as several others along the Delaware in the 1920s. The New York–Pennsylvania Joint Bridge Commission finally made a proposition to the officers of the Milanville Bridge Company that they apparently couldn't refuse. The Commission offered to buy this 470-foot, one-way bridge for $19,542.22, and it was promptly accepted. The bridge under its new ownership was immediately opened for use with the toll charges totally eliminated. Local citizens and visiting businessmen became active bridge users.

A regular maintenance program was initiated for the bridge, and this span has led a reasonably good life. Floods on the river in 1955 and again, later in the 1960s, seemed to be harmless at Skinner's Falls. In 1986, from May to October, the bridge was closed while major modernizing was done. Floor beams and trusses were replaced, new guide rails constructed, and the whole bridge repainted. And the approaches to the bridge were all repaved. The three-ton weight limit on the old bridge was the only thing not changed. While all this work was being done, the nearby competitive bridges were designated as official detours and got all the Milanville traffic until it was finished. Then in 1988, the updated bridge was added to the list of historic places on the National Register.

This honored and updated one-way bridge still serves travelers; the younger generation of users doesn't seem to mind waiting, occasionally, for their turn to cross the beautiful Delaware. If they aren't in a big hurry they can spend some time fishing, or just looking, at nearby Skinner's Falls. And the scenery in this area is truly wonderful to behold.

BIBLIOGRAPHY

BOOKS

Allen, Richard Sanders. *Covered Bridges of the Mid-Atlantic States.* Brattleboro, Vt.: Bonanza Books, 1959.

Atwood, Albert W. "Today on the Delaware: Penn's Glorious River." *National Geographic,* July 1952.

Barber, John W., and Henry Howe. *Historical Collections of New Jersey.* Trenton, 1844.

Child, Hamilton. *Sullivan County Gazetteer, 1872–1873.* Syracuse, N.Y.: Sullivan County Press, 1873.

Davis, William W. *History of Bucks County.* Volume 2. Genealogical Publishing Co., 1905.

Decker, Amelia Stickney. *Old Mine Road.* Sussex, N.J.: Wantage Recorder Press, 1932.

Delaware River Joint Bridge Commission, P.O. Box 88, Morrisville, Pa. 19067. All publications relating to Delaware River bridges.

"Eddies and Rifts." In *The Upper Delaware Story.* Narrowsburg, N.Y.: National Park Service, 1980.

Emerson, Harry. *Delaware Wilds.* Farrar & Rinehardt Press, 1940.

Fackenthal, B. F. *Improved Navigation on the Delaware River with Some Account of Its Ferries, Bridges, Canals, and Floods.* Bucks County Historical Society, 1927.

Fluhr, George J. *Pike in Pennsylvania: History of a County.* Lackawaxen, Pa.: Alpha Publishing Co., 1993.

———. *The Pond Eddy Bridge.* Shohola, Pa.: Privately printed, 1992.

Gardner, John Palmer. *The Valley of the Delaware.* Philadelphia: John C. Winsome Co., 1934.

Henn, William F. *The Story of River Road*. Library of Congress, Card No. 75-23404. Privately printed, 1992.

Hine, C. J. *The Old Mine Road*. 1909. Reprint, New Brunswick, N.J.: Rutgers University Press, 1963.

Hoff, J. Wallace. *Two-Hundred Miles on the Delaware River*. Trenton, N.J.: Brandt Press, 1893.

Jervis, John B. *The Reminiscences of John B. Jervis*. Edited by Neal Fitzsimmons. Syracuse, N.Y.: Syracuse University Press, 1971.

Johnston, John Willard. *Reminiscences*. Walton, N.Y.: Reporter Co., 1987.

Lee, Warren, and Catherine Lee. *A Chronology of the Belvidere-Delaware Railroad*. Albuquerque, N.M.: Sandia National Laboratories, 1989.

Letcher, Gary. *Canoeing the Delaware River*. New Brunswick, N.J.: Rutgers University Press, 1985.

Ludlum, David M. *The Weather Factor*. Boston: Houghton Mifflin, 1984.

McCaffrey, Katherine, and Marian Swope. *Bicentennial History of the Town of Lumberland*. N.Y.: Sullivan County Press, 1976.

Meyers, Arthur N. *Crossing the Delaware River . . . Via Toll Bridges*. Narrowsburg, N.Y.: Delaware Valley Press, 1970.

Pine, Joshua. *Rafting Story of the Delaware River*. Privately printed, 1873.

Plowdin, David. *Bridges*. New York: Macmillan, 1993.

Sanderson, Dorothy H. *The Delaware and Hudson Canal Way*. Ellenville, N.Y.: Round Valley Publication Co., 1965.

Schmidt, Robert. *Rural Hunterdon County*. New Brunswick, N.J.: Rutgers University Press, 1946.

Schwarz, Frank V. *Lumberland, A Gem With Many Facets*. Matamoras, Pa.: Williams Printing, 1998.

Shank, William. *Historic Bridges of Pennsylvania*. New York: American Canal Transportation Center, 1980.

Shaughnessey, James. *Delaware and Hudson*. Berkeley, Calif.: Howell-North Books, 1967.

Snell, James P. *History of Sussex, Warren, and Hunterdon Counties*. Philadelphia: Everts & Peck, 1881.

Stutz, Bruce. *People and Places on the River*. New York: Crown Publishing, 1992.

Trenton Historical Society. *The History of Trenton*. Princeton, N.J.: Trenton Historical Society, 1929.

Tyler, David. *The Bay and River Delaware*. Cambridge, Md.: Cornell Maritime Press, 1955.

Wakefield, Manville. *Canal Boats to Tidewater: Delaware and Lackawaxen Aqueducts*. Grahamsville, N.Y.: Wakefair Press, 1965.

Weiss, Harry B., and Grace M. Weiss. *Rafting on the Delaware River*. Trenton, N.J.: New Jersey Agricultural Society, 1967.

Wolfe, Nancy M. *The Area Guide Book: Bucks and Hunterdon Counties.* EnandEm Graphics Company, 1997.

Woods, Leslie. *Rafting on the Delaware River.* Livingston Manor, N.Y.: Livingston Manor Times, 1934.

Wynkoop, Ronald, Sr. *A Time to Remember.* Phillipsburg, N.J.: Sheraton Printing, 1985.

Zacher, Susan. *The Covered Bridges of Pennsylvania.* Harrisburg: Pennsylvania Historical and Museum Commission, 1982.

PERIODICALS

"Accounting of Rafts." *Port Jervis (New York) Daily Union,* April 20, 1880.

"Bridges Over the Delaware." *River Reporter* 25, no. 26. Narrowsburg, N.Y.

Clines, Francis X. "About Minisink Ford: Across the Delaware." *New York Times,* April 26, 1979.

"Delaware River Dam Destruction at Long Eddy." *Port Jervis (New York) Evening Gazette,* March 19, 1870.

"New Interstate Bridge." *Narrowsburg (New York) News-Times,* September 9, 1954.

"Old Iron Bridge Here Torn Down for Scrap." *Narrowsburg (New York) Times,* September 9, 1954.

INDEX

ABOUT THE AUTHOR

Frank T. Dale is a freelance writer and local historian. He has won awards from the New Jersey Society of Professional Journalists, the Working Press Association, and the New Jersey Historical Commission. His previous book, *Delaware Diary: Episodes in the Life of a River,* was published by Rutgers University Press in 1996. He and his wife currently reside in Hope, New Jersey.